Bioceramics

Applications of
Ceramic and Glass Materials
in Medicine

Editor:

James F. Shackelford

Department of Chemical Engineering and Materials Science
University of California
Davis, CA 95616

ttp **TRANS TECH PUBLICATIONS LTD**
Switzerland • Germany • UK • USA

Volume 293 of
Materials Science Forum
ISSN 0255-5476

Distributed in the Americas by

Trans Tech Publications Inc
PO Box 699, May Street
Enfield, New Hampshire 03748
USA

Phone: (603) 632-7377
Fax: (603) 632-5611
e-mail: ttp@ttp.net
Web: http://www.ttp.net

and worldwide by

Trans Tech Publications Ltd
Brandrain 6
CH-8707 Uetikon-Zuerich
Switzerland

Fax: +41 (1) 922 10 33
e-mail: ttp@ttp.ch
Web: http://www.ttp.ch

Printed in the United Kingdom
by Hobbs the Printers Ltd,
Totton, Hampshire SO40 3WX

Preface

This monograph on bioceramics was produced at the request of Dr. Fred Wöhlbier of Trans Tech Publications for the purpose of providing a compact, up-to-date reference on these important biomaterials. Engineered materials play a critical role in modern biomedical applications, and crystalline ceramics (and the chemically-related, noncrystalline glasses) are finding increasingly wide uses. I am especially grateful for the manuscripts which I solicited from some of the leading researchers on these materials. The authors have provided an admirable blend of specific research results and general overviews of advances in the field.

To begin the book, I provide a brief historical perspective on the application of ceramics and glasses to medicine. We find that the biomedical applications have largely developed during this century, with the "modern era" beginning in the 1960's. An important frame of reference in developing engineered materials for medical applications is the understanding of the nature of the natural materials that are being replaced by the synthetics. In this regard, Bruce Martin provides an especially interesting perspective on bone as a ceramic composite material. Tom McGee and co-workers then evaluate how a variety of biomaterials, including bioceramics, perform as synthetic tooth roots.

Larry Hench, the principal pioneer in using glass materials in biomedicine, provides a broad overview of their applications, along with a discussion of related glass-ceramics. Professor Hench's approach has consistently focused on the use of surface reactive materials. Tadashi Kokubo then illustrates numerous, recent developments which have grown out of Hench's approach to bioceramics. A highly impressive range of creative biomaterials have emerged from Japanese laboratories and especially from Professor Kokubo's group. Mehmet Sarikaya and co-workers give a contemporary overview of one of the most exciting areas of bioceramics, the development of biomimetic processing technologies. Finally, I close the monograph with a perspective on the current status and future trends in the use of bioceramics. Perhaps the most interesting result of the exercise in looking forward to bioceramics in the next century is to see the blurring of boundaries between the physical and biological sciences. This merger of ceramic science and engineering with modern biology is also illustrated repeatedly through the contributions from my colleagues earlier in the volume.

It is impossible for me to write on the subject of biomedical engineering without thanking my various collaborators over the years in the Orthopaedic Research Laboratory at the University of California, Davis Medical Center. It has been both my pleasure and good fortune to be associated with such an outstanding organization. I am also deeply grateful to Dr. Terence Mitchell and his colleagues at the Center for Materials Science at the Los Alamos National Laboratory for hosting my sabbatical leave which makes the completion of this project possible. Finally, I hope that the readers will find this monograph both useful and stimulating in relation to their own studies and applications of bioceramics.

J.F.S.
Davis, California

Table of Contents

Materials Science Forum Vol. 293 (1999) pp. 1-4
© 1999 Trans Tech Publications, Switzerland

Bioceramics - An Historical Perspective

J.F. Shackelford

Department of Chemical Engineering and Materials Science, University of California, Davis, CA 95616, USA

Keywords: Bioceramics, Biomaterials, Ceramics, Glasses

ABSTRACT

Bioceramics can be defined within the context of the broad family of engineering materials, as well as the *biomaterials*, i.e. those engineered materials used specifically for medical applications. Although not as widely used as metallic implants and various biopolymers, bioceramics have been the focus of close scrutiny for over three decades and are now finding some significant applications, especially in orthopaedics and dentistry. Bioceramics can be categorized by their degree of chemical reactivity. Oxide ceramics are often used for their relatively inert behavior. At the other extreme, resorbable bioceramics are highly reactive in relation to the physiological environment. Finally, surface reactive materials have intermediate reactivity and are designed to provide direct bonding to tissues.

THE CONTEXT OF BIOMATERIALS

Bioceramics are appropriately considered in the context of the broader topic of *biomaterials*. Any structural application for medical purposes makes the engineered material a "biomedical" one. As a result, biomaterials share with all engineering materials the classification into the traditional categories of *metals*, *ceramics* (and *glasses*), *polymers*, and *composites*. [1] An historical review of the field of biomaterials reveals that metallic and polymeric materials are generally more widely used in medical applications than are ceramics. To date, modern composites have been used in rather limited ways.

The broad history of biomaterials can be represented by the widely used metallic implants. [2] For example, metals have been used for orthopaedic applications since ancient times. Until 1875, relatively pure metals such as gold, silver, and copper were generally used. Unfortunately, poor surgical conditions limited their success. Engineered metal alloys became more widely used between

1875 and 1925 coincident with substantial improvements in surgical techniques. The period since 1925 can be considered the modern era of metallic biomaterials. A wide variety of orthopaedic applications have been developed for which the dominant alloys of choice are 316L stainless steel, Co-Cr alloys, and Ti-6Al-4V.

Polymeric biomaterials also have a long and rich history. Evidence exists in papyrus records for the use of linen sutures for closing wounds 4,000 years ago. Catgut was introduced for sutures in the second century. Silk was used for this purpose in the 11th century. A variety of contemporary synthetic polymers are now used in modern surgery. Polyethylene, polyester, polyglycolic acid, and nylon are examples. Biopolymers are currently used in a wide variety of surgical applications, including blood vessel prostheses, tissue adhesives, heart valves, lenses, and sutures. [3] In the widely used orthopaedic surgery of total hip replacement, polymethylmethacrylate (PMMA) cement and a polyethylene acetabular cup are routinely used.

As noted earlier, the use of *advanced composites* for biomedical applications has been the focus of much speculation but relatively limited use to date. [4] Carbon fiber-reinforced polymers have been used for structural applications such as the femoral stem in the total hip replacement. The beneficial feature of being able to control the stem modulus, however, is offset by concerns about physiological reactions to fibers which may be released into the biological environment. As a practical matter, the use of these and other "new" materials is especially challenging in the United States due to the relatively conservative policies of the Food and Drug Administration (FDA). The issue of governmental regulation can also be a factor in the expansion of applications for any biomaterial, including bioceramics.

THE CONTEXT OF CHEMICAL REACTIVITY

Orthopaedic bioceramics provide the advantage of chemical similarity to natural skeletal materials. Dental applications for ceramics are also attractive due to the chemical similarity between engineered ceramics and natural dental materials, as well as a predominance of compressive loads for which ceramics provide optimal mechanical performance. In contrast, the mechanical loading for orthopaedic applications tends to include substantial tensile stress components. Three broad categories of bioceramics (based on chemical reactivity with the physiological environment) have been defined by Hulbert, et. al. [5] as summarized in Table 1 and illustrated by figure 1.

Table 1. Categories of Ceramic Biomaterials[1]

Category	Example
Inert	Al_2O_3
Surface Reactive	Bioglass
Resorbable	$Ca_3(PO_4)_2$

[1]After Hulbert, et.al. [5]

Relatively *inert* bioceramics, such as structural Al_2O_3, tend to exhibit inherently low levels of reactivity which peak on the order of 10^4 days (over 250 years). *Surface reactive* bioceramics, such as Hench's Bioglass, [6] have a substantially higher level of reactivity peaking on the order of 100 days. *Resorbable* bioceramics, such as tricalcium phosphate, have even higher levels of reactivity peaking on the order of 10 days. This broad spectrum of chemical behavior leads to a corresponding range of engineering design philosophies. Tricalcium phosphate is representative of a resorbable bioceramic. The oxide ceramics represent the opposite strategy, viz. a nearly inert bioceramic. The intermediate approach using surface reactive bioceramics was developed by Hench and co-workers in the early 1970s. [6-10] Their primary development was *Bioglass*, defined as a glass designed to bond directly to bone by providing surface reactive silica, calcium, and phosphate groups in an alkaline pH environment. Bioglass can be alternatively defined as a soda-lime-silica glass with a significant phosphorous oxide addition, e.g. Bioglass 45S5 containing 45 wt% SiO_2, 24.5 wt% CaO, 24.5 wt% Na_2O, and 6 wt% P_2O_5. This material is noticeably lower in silica and higher in lime and soda than conventional window and container glasses. The practical application of Bioglass in orthopaedics has been limited, largely due to the slow kinetics of surface reaction rates and the corresponding slow development of interfacial bond strength. Roughly 6 months are required before the interfacial strength approaches that provided by traditional *polymethylmethacrylate (PMMA)* cement after 10 minutes setting time. On the other hand, Bioglass and related materials have found wide uses in dentistry and ear surgeries.

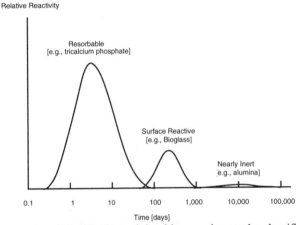

Figure 1 In the context of chemical reactivity, bioceramics can be classified into three subgroups, based on their reactivity with the physiological environment. (After Hulbert, et.al. [5])

THE DEVELOPMENT OF BIOCERAMICS

In turning our attention to the historical development of applications in biomedicine for ceramics and glasses, it is important to maintain a perspective on the related applications of the other structural materials reviewed above. The inherent brittleness of traditional ceramics has generally limited their competition with ductile metals and polymers. This is offset by the obvious fact that bone is 43% by weight hydroxyapatite, a common ceramic mineral. Current advances in ceramic processing, including the production of significant improvements in fracture toughness, are contributing to increased possibilities for ceramics in biomedicine.

The history of bioceramics has been reviewed in some detail by Hulbert, et. al. in the early 1980s. [5] The first widely evaluated bioceramic was common plaster of Paris, $CaSO_4 \cdot H_2O$. Dreesman published the first report on its use to repair bone defects in 1892. [5] Plaster of Paris continued to be extensively studied for such applications through the 1950's. [11] This pioneering bioceramic represents the characteristic trade off between good physiochemical behavior and poor mechanical performance. Attractive features of plaster of Paris included little or no adverse tissue reaction and its replacement by new tissue at a rate comparable to its absorption by the physiological system. These advantages were unfortunately offset by an inherent weakness and a rapid degradation in strength during absorption.

As early as 1920, a successful application of *tricalcium phosphate*, $Ca_3(PO_4)_2$, was reported. [12] The average length of time for bone defect repair in rabbits was found to be accelerated from 41 days to 31 days. In contrast, an unsuccessful application of a calcium salt was demonstrated by numerous studies on calcium hydroxide, which tends to stimulate the formation of immature bone. [11]

The "modern" era of bioceramics is generally traced to Smith's successful 1963 study on Cerosium, a ceramic bone substitute composed of a porous aluminate ceramic impregnated with an epoxy resin. [13] The porosity of the ceramic was controlled at 48% to duplicate the value for natural bone and to produce net physical properties very close to those of bone. Similar modulus and flexural strengths combined with good biocompatibility led to successful bone replacement applications during the remainder of the 1960's and into the 1970's. During this time, a widespread interest in biomedicine developed within the ceramics community, largely as a result of the extensive work of Hulbert and co-workers. [14-17] They demonstrated convincingly the biocompatibility of *oxide ceramics*, while developing the use of bone tissue ingrowth into porous ceramics as a means for mechanically interlocking prostheses. As noted above, Hench and co-workers [6] pioneered the study of surface reactive materials such as Bioglass in the early 1970s.

The interest in *bioceramics* plateaued by the end of the 1970s but has increased substantially in recent years, primarily in orthopaedics and dentistry. [18-20] Much of this renaissance has been associated with the successful applications of hydroxyapatite (HA) ceramics. Examples include an HA coating which facilitates the fixation of the total hip replacement prosthesis [21] and an HA-based composite for repair of large bone defects. [22]

References

1) Shackelford, J.F.: *Introduction to Materials Science for Engineers*, 4th Edition, 1996, Prentice-Hall, Upper Saddle River, NJ.
2) Fraker, A.C. and Ruff, A.W.: *J. Metals,* 1977, **29,** 22.
3) Ratner, B.D.: *J.Biomed.Mater.Res.,* 1993, **27, 837**.
4) Devanathan, D.: p. 74 in *International Encyclopedia of Composites*, Vol. 4, ed. Lee, S.M.: 1991, VCH Publishers, New York.
5) Hulbert, S.F., Hench, L.L., Forbers, D., and Bowman, L.S.: *Ceramurgia Intl.,* 1982-83, **8-9,** 131.
6) Hench, L.L., Splinter, R.J., Allen, W.C., and Greenlee, Jr., T.K.: *J.Biomed.Mater.Res.,* 1971, **5,** 117.
7) Hench, L.L. and Paschall, H.A.: *J.Biomed.Mater.Res.,* 1973, *7,* 25.
8) Piotrowski, G., Hench, L.L., Allen, W.C., and Miller, G.J.: *J.Biomed.Mater.Res.,* 1975, **9,** 47.
9) Griss, P., Greenspan, D.C., Heimke, G., Krempien, B., Buchinger, R., Hench, L.L., and Jentschura, G.: *J.Biomed.Mater.Res.,* 1976, **10,** 511.
10) Stanley, H.R., Hench, L.L., Going, G., Bennett, G., Chellemi, S.J., King, C., Ingersoll, N., Ethridge, E., and Kreutziger, K.: *Oral.Surg., Oral Med., Oral Path.,* 1976, **42,** 339.
11) Peltier, L.F., Bickel, E.Y., Lillo, R., and Thein, M.S.: *Ann.Surg.,* 1957, **146, 61**.
12) Albee, F.H. and Morrison, H.F.: *Ann.Surg.,* 1920, **71,** 32.
13) Smith, L.: *Arch.Surg.,* 1963, **87,** 653.
14) Hulbert, S.F., Young, F.A., Mathews, R.S., Klawitter, J.J., Talbert, C.D., and Stelling, F.H.: *J.Biomed.Mater.Res.,* 1970, **4,** 433.
15) Hulbert, S.H.: in: *1st Intl. Biomaterials Symp.,* 1969, Clemson, South Carolina.
16) Talbert, C.D.: Masters Thesis, 1969, Clemson University, Clemson, South Carolina.
17) Klawitter, J.J.: Ph.D. Thesis, 1970, Clemson University, Clemson, South Carolina.
18) *Bioceramics: Proc. 4th Intl. Symposium on Ceramics in Medicine, London, 1991*, ed. Bonfield, W. , Hastings, G.W., and Tanner, K.E.: 1991, Butterworth-Heinemann, Oxford, England.
19) *Bioceramics and the Human Body, Proc. Symposium, Faenza, Italy, 1991*, ed. Ravaglioli, A. and Krajewski, A: 1992, Elsevier Applied Science, New York.
20) *Bioceramics: Materials and Applications, Proc. Symposia at Indianapolis, Indiana and Alfred, NY, 1994*, ed. Fishman, G.S., Clare, A., and Hench, L.L.: 1995, American Ceramic Society, Westerville, Ohio.
21) de Lange, G.L. and Donath, K.: *Biomaterials,* 1989, **10, 121**.
22) McIntyre, J.P., Shackelford, J.F., Chapman, M.W., and Pool, R.R.: *Am.Ceram.Soc.Bull.,* 1991, **70** 1499.

Materials Science Forum Vol. 293 (1999) pp. 5-16
© 1999 Trans Tech Publications, Switzerland

Bone as a Ceramic Composite Material

R.B. Martin

Orthopaedic Research Laboratories, University of California,
Davis Medical Center, Sacramento, CA 95817, USA

Keywords: Bone, Structure, Mineralization, Porosity, Mechanical Properties

ABSTRACT
Bone is the primary structural material used to carry major loads in an enormous variety of vertebrate animals. To fulfill this role, nature has devised an extremely interesting ceramic composite whose components are primarily collagen and hydroxyapatite, but whose complex structure contains a wealth of mechanically relevant detail. It is a composite in several different senses, being a porous material, a polymer-ceramic mixture, a lamellar material, and a fiber-matrix material. Its mechanical properties depend on each of these aspects of composition and structure. Variations in porosity and mineralization serve to adapt bone's mechanical properties to suit a great variety of conditions. Lamellar and fiber-matrix interfaces limit the growth of cracks and fatigue damage. Through the actions of remodeling by cells, bone's composite structure repairs fatigue damage and continuously adapts to changes in mechanical usage.

INTRODUCTION
Plants and animals have evolved a great variety of structural materials to serve their needs. Many of these have been usurped by humans as engineering materials, wood being the most obvious example in modern times. In pre-historic times, tendon, animal skins, and plant fibers were also widely used in the construction of many tools and structures, but bone was perhaps the most important biologically-derived *structural* material to be used, other than wood. The properties which made it useful for the construction of tools are those vital to its original role in the skeleton: strength, toughness, fatigue resistance, and lightness. This combination of properties is derived from the fact

that bone is, to a much greater degree than most other biological materials, a ceramic composite. That is, it is composed of an organic polymer and a mineral in fairly equal proportions, not just mixed together, but with an integrated structure which is optimized for mechanical function. The purpose of this chapter is to explore this material and its mechanical properties.

THE STRUCTURE OF BONE

Bone is a composite material in several different ways. First of all, it is porous material composed of a solid matrix containing voids of varying sizes which are filled with soft tissues. These soft tissues include various bone cells and their blood supply, as well as marrow, which is an entirely different kind of tissue than bone,[1] but one which has an intimate physiological relationship to the bone cells. While the porosity of bone varies continuously from about 5% to about 95%, the porosity distribution is bi-modal, with most bone being either "compact" (of low porosity) or "spongy" (of high porosity). (It will be seen that variations in porosity are the primary means of controlling bone's mechanical properties.) Figures 1 and 2 illustrate the structures of compact and spongy bone. Compact bone is found in the shafts of long bones and as a shell surrounding spongy bone regions. Spongy bone is found within the ends of long bones, inside the vertebral bones of the spine, and sandwiched between layers of compact bone in the skeleton's plate structures (e.g., the pelvis and skull).

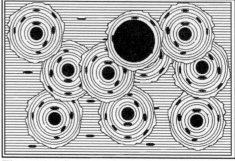

Figure 1: Cross-section of compact bone showing Haversian canals in osteons on background of primary bone.

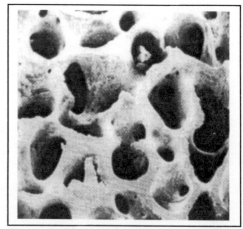

Figure 2: Picture of spongy bone showing trabecular plates 100-200 microns thick. From ref. (26) with permission.

The second way in which bone is a composite material stems from its lamellar structure. Most compact bone has a plywood-like structure, with each lamella being about 5 μm thick. Bone is formed by cells called *osteoblasts*. Typically, the initial bone to be laid down has lamellae parallel to the bone's outer surfaces (*primary bone* in the background of figure 1). Embedded in the bone are cells called *osteocytes*, which are actually osteoblasts which became trapped in the bone they were forming. The osteocytes live in small chambers called *lacunae* (the small inclusions in figure 1) which are connected by *canaliculi* (literally, "little canals," not shown). The lacunae and canaliculi are 5-10 and about 0.2 microns in diameter, respectively [6,29], and constitute the smallest voids in the bone as visualized by light or

---

[1] The function of marrow is to manufacture blood cells.

electron microscopy. It may also be noted that osteocytes are thought to be important in the mineralization process.

Subsequently, structures called *osteons* are created inside the primary bone. These are tubular structures some 200 μm in diameter, with concentric lamellae arranged about a central *Haversian canal* containing a blood vessel. These are formed by cells called *osteoclasts*, which dig a tunnel through the solid matrix, parallel to the bone shaft's axis. Osteoblasts refill the tunnel to the 50 μm diameter of the Haversian canal. This process is called *remodeling*. The Haversian canals and the active remodeling tunnels constitute the largest voids in compact bone. Early in life osteons come to dominate the bone structure, so that little primary bone exists and one has a third kind of composite structure consisting of many tubular "fibers" (the osteons) in a matrix comprised of old osteon fragments. This fiber-matrix composite material is called *secondary* or *osteonal* bone. To reiterate, compact bone contains three constituent types of composite structure: a microscopically porous composite, a lamellar composite, and a fiber-matrix composite. The overall architecture which determines compact bone's mechanical properties is primarily the orientation, size, and packing of the osteons and their Haversian canals.

Spongy bone is also called *trabecular* or *cancellous* bone. In it the minor void spaces are composed of the same lacunar-canalicular network as in compact bone. Now, however, the porous lamellar matrix is formed into plates or struts (*trabeculae*) approximately 100 μm thick (figure 2). These structures are too thin to contain an osteon, so there are no Haversian canals. Instead, the major porosity component is the connected network of marrow-filled voids surrounding the trabeculae. The characteristic dimension of these voids is not much greater than that of the larger voids in compact bone, but their total volume fraction is much greater, as previously noted. Even though osteons cannot form inside trabeculae, remodeling still occurs by digging and refilling trenches on trabecular surfaces, rather than tunneling through the trabeculae. Consequently, the bone matrix consists of a patchwork of small regions with parallel lamellae, like the inter-osteonal fragments in compact bone. Thus, spongy bone is, like compact bone, a composite of three constituent types, with the third level being different: a microscopically porous composite, a lamellar composite, and a macroscopically porous, foam-like composite. The overall architecture which determines spongy bone's mechanical properties is primarily the thickness, connectivity, and orientation of the trabeculae, rather than the organization of the bone tissue inside.

THE COMPOSITION OF BONE

The composition of the solid portion of bone may be broken into three major components: an organic polymer, a ceramic, and viscous liquids. The polymer is collagen, a protein which is the most abundant structural material in the bodies of mammals. There are actually more than a dozen different kinds of collagen, each determined by the particular sequence of amino acids in its molecules.

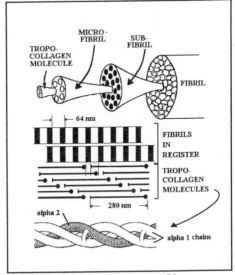

Figure 3: Collagen structure of bone.

The collagen in bone is Type I, the same as is found in skin, tendons, and ligaments. Figure 3 shows how the molecules of Type I collagen are formed into fibrils containing the characteristic 64 nm banding pattern.   The fact that this banding pattern is in register across different fibrils is an indication that the fibrils are not mechanically independent of one another, but are connected by external as well as internal molecular cross-linking. In bone these collagen fibrils are organized into fibers, which in turn are laid down parallel to one another to form the lamellae shown in figure 1. There is some evidence that the fiber orientations alternate from lamella to lamella, as in a cross-ply laminate, but in fact the details of this structure are not clear [15]. The organic portion of the bone matrix also contains a small amount of other proteins, and molecules called mucopolysaccharides, but their mechanical roles are also in question.

The ceramic component of bone is comprised of a variety of calcium phosphates. All but a small percentage of this is essentially hydroxyapatite, $Ca_{10}(HPO_4)_6(OH)_2$, but there are many substitutions in its molecules: K, Mg, Sr, and Na for Ca; carbonate for phosphate, and fluorine for OH [26]. Some of these substitutions are thought to play significant roles in the structure and mechanical properties of the bone mineral. (For example, excessive fluoride from drinking water may weaken bone via this effect, as well as by harming bone cells.) It is thought that hydroxyapatite (HA) cannot be directly precipitated from the extracellular fluids in bone, but that precursor calcium phosphates must be formed first, and transformed into HA [1]. The putative initial mineral is brushite. These precursor mineral forms comprise the balance of bone's ceramic phase.

The relationships between bone's collagen and mineral phases are thought to be manifold and complex. When osteoblasts make bone, it contains water in the place of mineral. Gradually, the organic matrix (called *osteoid*) "matures" and calcification begins. The initial precipitation of mineral crystals is postulated to depend on catalysis by the elements of the collagen structure, as well as other biochemical regulators [1]. Then the physical arrangement of the two phases develops based on the structure of each. The initial crystals grow in the collagen bands, then spread throughout the collagen scaffold. In humans, new bone reaches about 70% of its mineral capacity over a period of roughly 4 days. This process is known as *primary mineralization* (figure 4). *Secondary mineralization* requires

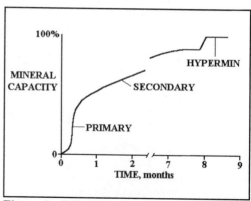

Figure 4: Schematic diagram of changes in mineralization of new bone over time.

months to slowly and asymptotically bring the bone to a "normal" mineral content [20]. However, this is not the maximum mineralization that the bone can achieve. Sometimes the bone matrix is observed to be *hypercalcified* [7]. (In figure 4 this is arbitrarily depicted as having occurred about 8 months after the initial formation of a bone moiety, but it could happen years later.) Hypermineralization involves not only further mineralization of the organic matrix, but also mineralization of the soft tissues in the cell spaces. This is regarded as a pathologic condition

---

[2] This discussion of hypermineralization applies only to compact bone. The author is not aware of similar observations in spongy bone.

because it is not seen in "normal" bone, involves cell death, and would presumably make the bone more brittle and less tough. Hypermineralization has also been observed in association with fatigue damage [27]. It has been proposed that osteocytes maintain the mineral ion composition of bone fluids at a level below that of blood serum so that "normal" mineral content is held below that possible if the matrix were exposed to serum [7]. The death of osteocytes would then theoretically allow hypermineralization to occur.

THE MECHANICAL PROPERTIES OF BONE

Bone is a rather linear and brittle, yet anisotropic and viscoelastic material whose mechanical properties are determine by its porosity, degree of mineralization, collagen fiber orientation, and other structural details. Table 1 compares compact bone's density and elastic modulus to those of a variety of other materials, and ranks them according the ratio of modulus to density. While bone may be a wonderful material in the body of an animal, it falls dead last in Table 1. Of course, this table ignores the ability of bone to repair and remodel itself.

Table 1: Density and elastic modulus for a variety of materials, including bone.

MATERIAL	DENSITY, gm/ml	ELASTIC MOD., GPa	MODULUS/ DENSITY
Graphite fiber	1.8	276	153.3
SiF fiber, carbon core	3.0	406	135.3
Al_2O_3 fiber	3.9	385	98.7
Alumina	3.9	345	88.5
Hydroxyapatite	3.2	279	87.2
Hard woods (e.g., ebony)	1.3	100	76.9
HPZ fiber[a]	2.35	175	74.5
Ivory	1.9	90	47.4
Quartz	2.65	103	38.9
Polycrystalline glass	2.5	88	35.2
Aluminum	2.7	70	25.9
Stainless steel	8.02	193	24.1
Titanium	5.0	114	22.8
Marble	2.72	55	20.2
Granite	2.67	50	18.7
Zirconium	6.5	83	12.8
Compact bone	2.1	20	9.5

[a] A Si, C, N, O composite manufactured by Dow Corning/Celanese.

Table 2: Mechanical Properties of Human and Bovine Osteonal and Primary Bone

Elastic modulus, GPa	TENSION		COMPRESSION	
	Longitudinal	Transverse	Longitudinal	Transverse
Human osteonal	17.9 ± 9	10.1 ± 2.4	18.2 ± 0.9	11.7 ± 1.0
Bovine osteonal	23.1 ± 3.2	10.4 ± 1.6	22.3 ± 4.6	10.1 ± 1.8
Bovine primary	26.5 ± 5.4	11.0 ± 0.2	-------	-------
Ultimate stress, MPa	TENSION		COMPRESSION	
	Longitudinal	Transverse	Longitudinal	Transverse
Human osteonal	135 ± 16	53 ± 11	105 ± 17	131 ± 21
Bovine osteonal	150 ± 11[a]	49 ± 7[a]	272 ± 3	146 ± 32
Bovine primary	167 ± 9	55 ± 9[a]	-------	-------

Mean ± standard deviation (SD). Data from ref. [23].
[a] SD approximate due to group averaging.

Figure 5 shows typical stress-strain curves for compact and spongy bone [12]. Table 2 gives modulus and strength data for compact bone under several measurement conditions. The modulus of elasticity of compact bone is about 20 MPa., less than 1/3 that of aluminum but 7 times greater than acrylic plastics. Table 2 also shows the considerable anisotropy of compact bone. As remodeling replaces primary bone with osteons, the bone changes from an orthotropic material to one that is transversely isotropic. This is illustrated by the elastic coefficients shown in Table 3.

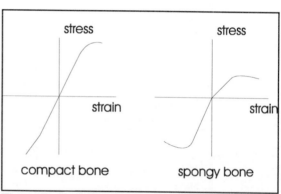

Figure 5: Typical stress strain curves for compact (left) and spongy (right) bone, to different scales.

Table 3: Elastic Constants of Primary (Orthotropic) and Osteonal (Transversely Isotropic) Bone, GPa

Type bone	c_{11}	c_{22}	c_{33}	c_{44}	c_{55}	c_{66}	c_{12}	c_{23}	c_{13}
Primary	22.4	25.0	35.0	8.2	7.1	6.1	14.0	13.6	15.8
Osteonal	21.2	21.0	29.0	6.3	6.3	5.4	11.7	11.7	12.7

Data from ref. [11].

DETERMINANTS OF BONE'S MECHANICAL PROPERTIES

Individually, compact and spongy bone exhibit rather linear relationships between modulus or strength and porosity, but when such a relationship is plotted for bone of all porosities, the graph has the characteristic hyperbolic shape seen in other porous materials (figure 6). Small changes in porosity affect the modulus or strength of compact bone proportionately more than similar changes in spongy bone. The modulus of spongy bone varies from about 10% that of compact bone down to virtually zero, and its compressive strength is less than 50 MPa [26]. Empirical relationships for the effect of porosity (p, a dimensionless volume fraction) on elastic modulus (in GPa) include, for bovine cortical bone [25],

$$E = 3.66 \ p^{-0.55} \qquad (1)$$

for human skull bone [17],

$$E = 12.4 \ (1 - p)^3 \qquad (2)$$

and for cortical bone from many species [5],

$$E = 3.94 \ (1 - p)^{5.74} \qquad (3)$$

Martin [13] has suggested

$$E = 15 \ (1 - p)^3 \qquad (4)$$

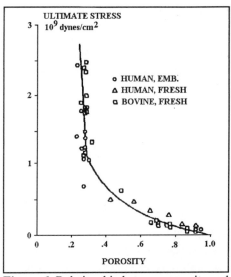

Figure 6: Relationship between porosity and elastic modulus for bone. From ref. [15], with permission.

as a useful approximation for cortical or spongy bone.

Bone's mechanical properties are even more sensitive to changes in mineralization than to porosity variability, but the effects of this sensitivity are damped by the fact that bone's mineral content does not vary nearly as much as its porosity. Usually, the mineral component of bone is 65-70% of the dry weight of the solid matrix, and rarely lies outside the 60-75% range. Currey (1988) has studied this aspect of bone's mechanical properties extensively, and has found that the elastic modulus varies with mineralization raised to a power. For example, in one study involving compact bone from a wide variety of species,

$$E = 1.48 \times 10^{-9} \ Ca^{4.15} \qquad (5)$$

where E is modulus in GPa and Ca is calcium content (mg Ca/gm of dry bone matrix). This equation predicts that a 4% change in mineral content would produce a 17% change in elastic modulus.

Katz [10] used mixture theory to study the feasibility of modeling bone as a amalgamation of collagen fibers and hydroxyapatite crystals. He assumed that the elastic moduli of collagen and hydroxyapatite were E_C and E_M, respectively, and that their volume fractions were V_C and V_M, respectively. (The void spaces in bone were ignored.) Katz found that the elastic modulus predicted by a Voigt (collagen fibers in parallel with mineral crystals, or uniform strain) model,

$$E_V = E_M V_M + E_C V_C \tag{6}$$

was substantially larger than that of bone, and the modulus predicted by a Reuss (fibers and crystals in series, or uniform stress) model,

$$E_R = E_M E_C / (E_C V_M + E_M V_C) \tag{7}$$

was much less than experimental values. Thus, these models only provide upper and lower bounds on the behavior, each of which was well outside the range of normal variation. For example, for an HA volume fraction of 40%, the Voigt and Reuss moduli were about 2 and 40 GPa, respectively. The data on bone with this mineral content were in the range of 15-25 GPa, so the models were not very helpful. Katz also found that the Hashin and Hashin-Shtrikman models for mixtures in which one phase is contained in inclusions provided somewhat better upper and lower bounds, but still did not serve as useful predictors of the elastic behavior of bone (or other mineralized tissues). It seems clear that an effective analytical solution to this problem must account for the structure as well as the relative volumes of the constituent materials. Unfortunately, no more recent models have appeared to solve this problem.

Taken together, the porosity and mineralization of bone determine its apparent density, i.e., its density including the soft tissues in the void spaces. Carter and Hayes [3] found that when both compact and spongy bone were included, the elastic modulus and compressive strength of bone depended on density cubed and squared, respectively:

$$E = 3790 \, r^{0.06} \, \rho^3 \tag{8}$$

$$\sigma_C = 68 \, r^{0.06} \, \rho^2 \tag{9}$$

Here, E is in GPa, σ_C is in MPa, and r is strain rate in sec^{-1}. Although the rigor of these equations has been questioned [24], they are useful approximations and have been widely accepted.

In addition to these compositional factors, bone's mechanical properties have been found to depend on several variables related to internal structure. For example, there are studies showing that more longitudinally arranged collagen fibers increase the tensile or bending strength and stiffness of both individual osteons and whole bone [14,16]. These observations are in concert with others showing that portions of the human femur which are habitually on the tensile side of the neutral axis for bending have more longitudinally oriented collagen [22]. Also, osteonal compact bone is weaker and less stiff than primary bone (table 2).

One may ask, if the production of osteons by remodeling weakens bone, what benefit does remodeling provide? The answer seems to be related to fatigue and fracture resistance. Osteons, as well as bone's overall lamellar structure, afford the same protection against crack propagation as is seen in engineering composite materials. The interfaces between lamellae and osteonal cement lines "capture" cracks and divert them into longitudinal directions, dissipating energy and preventing the growth of transverse cracks which would cause fracture. Experimental data show the osteonal pullout increases the fracture toughness of bone [21] and is associated with improved fatigue resistance [27]. In addition, remodeling removes bone bearing fatigue damage and replaces it with undamaged material.

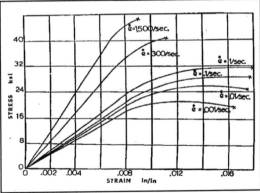

Bone is viscoelastic, exhibiting creep and relaxation effects. McElhaney's [18] classic paper showed that increased strain rates diminish the ductility but increase the modulus and strength of compact bone loaded in compression (figure 7). Interestingly, the work-to-failure increased with strain rate to a point, then declined. The peak of this curve occurred at about 1 sec.[-1], equivalent to

Figure 7: Effects of strain rate on human femur bone stress-strain curves. From ref. [18] with permission.

loading by a fall from a height, but less than strain rates produced by modern instruments of trauma, such as automobile collisions and bullets.

THE MECHANICAL ADAPTABILITY OF BONE

The bone mass in the pelvis of a paraplegic is reduced about 33% within six months of becoming immobilized [19]. Conversely, professional rodeo riders and tennis players have larger bones in the arms which experience extraordinary loading due to their vocations [4,9]. If a screw hole is drilled in the wall of a long bone cylinder, its work to failure in torsion is reduced 50-75%, but this deficiency corrects itself in about two months [2]. Unlike engineering materials, bone has the ability to alter its internal and external structures to adapt to changing mechanical requirements and minimize its weight in relationship to loading. It also has the ability to repair fatigue damage (as well as a complete fracture, of course, but that is beyond the scope of our consideration here). Indeed, the repair of fatigue damage and adaptation to changing mechanical conditions are essential functions of bone remodeling. In orthopaedics, this principle is called Wolff's Law, after a late nineteenth century observer [30].[1] Since this chapter deals primarily with bone as a ceramic composite *material*, only the adaptability of bone's internal structure will be considered here.

Remodeling changes the material properties of compact and spongy bone in several ways. First and foremost, remodeling changes porosity, and in two ways. Initially, there is a transient porosity change due to the fact that the first stage of remodeling is resorption of a cavity which is subsequently refilled. This transient porosity increase is called the *remodeling space*, and will persist so long as the remodeling rate remains elevated, then gradually decline as the additional resorption cavities are refilled. Remodeling may also permanently elevate porosity if refilling is inhibited. This is what happens in disuse osteoporosis in order to reduce bone mass.

Remodeling also changes the mineral content of the bone. If the remodeling rate is high, a high proportion of the bone matrix will be recently formed and therefore not fully mineralized. This is why children's bones are more compliant than those of adults. It is not that mineralization *per se* is inhibited in children. Rather, the remodeling rate is 1-2 orders of magnitude higher in their skeletons [15]. Finally, remodeling changes the organization of the calcified matrix. In compact bone it creates new osteons with lamellar structures oriented so as to reduce the stresses produced by the current bone geometry and applied loads. In spongy bone, it rearranges the trabecular architecture for the same purpose.

SUMMARY
Other biological composites - such as enamel and dentin - have evolved for more specialized purposes, but, at least in vertebrates, bone is nature's all-purpose, use-it-everywhere structural material. It has supported loads while minimizing weight in every sort of creature, from dinosaurs to cheetahs to hummingbirds. While giving a composite the ability to repair fatigue damage and adapt to changing loads will probably be an unreasonable design expectation for the foreseeable future, engineers may still hope to learn a trick or two from further study of nature's premier building material.

REFERENCES
1) Boskey, A. L., & Posner, A. S.: Orthopaedic Clinics of North America, 1984, 15, 597-612.
2) Burstein, A. H., Currey, J., Frankel, V. H., Lunseth, P., & Vessely, J. C.: The Journal of Bone and Joint Surgery, 1972, 54A (6), 1143-1156.
3) Carter, D. R., & Hayes, W. C.: Science, 1976, 194, 1174-1176.
4) Claussen, B. F.: Clinical Orthopaedics and Related Research, 1982, 164, 45-47.
5) Currey, J. D.: Journal of Biomechanics, 1988, 21, 131-139.
6) Frost, H. M.: Henry Ford Hospital Medical Bulletin, 1960, 8, 208-211.
7) Frost, H. M.: Journal of Bone and Joint Surgery, 1960, 42A, 144-150.
8) Hodgeman, C. D., & Lange, N. A.: Handbook of Chemistry and Physics, 1926, Cleveland, Chemical Rubber Publishing Co.
9) Jones, H. H., Priest, J. D., Hayes, W. C., Tichenor, C. C., & Nagel, D. A.: The Journal of Bone and Joint Surgery, 1977, 59-A, 204-208.
10) Katz, J. L.: Journal of Biomechanics, 1971, 4, 455-473.
11) Katz, J. L., Yoon, H. S., Maharidge, R., Meunier, A., & Christel, P.: Calcified Tissue International, 1984, 36, S31-S36.
12) Keaveny, T. M., Guo, X. E., Wachtel, E. F., McMahon, T. A., & Hayes, W. C.: Journal of Biomechanics, 1994, 27, 1137-1146.
13) Martin, R. B.: Journal of Biomechanics, 1991, 24, 79-88.
14) Martin, R. B., & Boardman, D. L.: Journal of Biomechanics, 1992, 26, 1047-1054.
15) Martin, R. B., & Burr, D. B.: The Structure, Function, and Adaptation of Compact Bone, 1989, New York, Raven Press.
16) Martin, R. B., & Ishida, J.: Journal of Biomechanics, 1989, 22, 419-426.
17) McElhaney, J., Alem, N., & Roberts, V.: ASME Paper No. 70-WA/BHF-2, 1970, 1-9.
18) McElhaney, J. H.: Journal of Applied Physiology, 1966, 21, 1231-1236.
19) Minaire, P., Meunier, P., Edouard, C., Bernard, J., Courpron, P., & Bourret, J.: Calcified Tissue Research, 1974, 17, 57-73.
20) Parfitt, A. M.: The physiologic and clinical significance of bone histomorphometric data, 1983, in R. R. Recker (Eds.), Bone Histomorphometry Techniques and Interpretation, Boca Raton, CRC Press, Inc., 143-223.
21) Piekarski, K.: Journal of Applied Physics, 1970, 41, 215-223.
22) Portigliatti-Barbos, M., Bianco, P., Ascenzi, A., & Boyde, A.: Metabolic Bone Disease and Related Research, 1984, 5, 309-315.
23) Reilly, D. T., Burstein, A. H., & Frankel, V. H.: Journal of Biomechanics, 1974, 7, 271-275.
24) Rice, J. C., Cowin, S. C., & Bowman, J. A.: Journal of Biomechanics, 1988, 21, 155-168.

25) Schaffler, M. B., & Burr, D. B.: Journal of Biomechanics, 1988, 21, 13-16.

26) Simon, S. R. (Ed.): Orthopaedic Basic Science, 1994, Rosemont, IL, American Academy of Orthopaedic Surgeons.

27) Stover, S. M., personal communication, 1994.

28) Weast, R. C. (Ed.): CRC Handbook of Chemistry and Physics, 1983, Boca Raton, FL, CRC Press, Inc.

29) Weinbaum, S., Cowin, S. C., & Zeng, Y.: Journal of Biomechanics, 1994, 27, 339-360.

30) Wolff, J.: Das Gasetz der Transformation der Knochen, 1892, Berlin, Hirschwald.

Materials Science Forum Vol. 293 (1999) pp. 17-36
© 1999 Trans Tech Publications, Switzerland

Evaluation of the Tissue Response of Organic, Metallic, Ceramic and Osteoceramic Tooth Roots

K.S. Tweden[1], G.I. Maze[2], T.D. McGee[3], C.L. Runyon[4] and Y. Niyo[3]

[1] St. Jude Medical, Inc., One Lillehei Plaza, St. Paul, MN 55117, USA

[2] College of Dentistry, University of Nebraska Medical Center,
40th and Holdrege, Lincoln, NE 68583-0740, USA

[3] Biomedical Engineering Program, Iowa State University, Ames, Iowa 50011, USA

[4] Department of Companion Animals, Atlantic Veterinary College, University of Prince Edward
Island, 550 University Avenue, Charlottetown, P.E.I., Canada C1A493

Keywords: Tooth Roots, Osteoceramic, Pyrolyte, Bioceram, Core-Vent, Tissue Response

ABSTRACT

Four classes of materials, inert organic, inert metal, inert ceramic and biologically-active osteo-ceramic were implanted in the edentulous mandibles of dogs. Tissue response was evaluated at 3, 6, 9, 12 and 18 months by light microscopy; microradiography; clinical evaluation for mobility, rotation, bleeding and radiography; and epi fluorescent analysis of bone growth labels. The pyrolytic carbon implants were not osseous integrated. The bone contact area of the sapphire and titanium alloy implants increased to 60 and 80%, respectively, at 12 months; but decreased to 20% and 60% at 18 months. The area for the osteoceramic was 80% at three months and remained at that level. These results indicate superior tissue response for the osteoceramic material, a ceramic composite of $Ca_3(PO_4)_2$ and $MgAl_2O_4$.

INTRODUCTION

Placement of endosseous dental implants has increased nearly two fold since the introduction of titanium endosseous root form implants in 1986 (Stillman and Douglas, 1993). Presently more than 400,000 dental implants are placed annually. However, demands from 40 million fully edentulous and many more partially edentulous individuals will dramatically increase this number with continued success of these devices (Schnitman, 1993). Eighty eight percent of implants used today are a root form shape and made of pure titanium, titanium alloy or a hydroxylapatite coated titanium. Pure titanium cylindrical screw implants are highly successful showing 5-8 year survival rates of 99.1% for the mandible and 85% for the maxillae (Albrektsson et al. 1988). Early carbon implants were vitreous carbon or pyrolytic carbon, the latter being commercially available to this study (Meffert 1985).

Titanium implants coated with hydroxylapatite show greater bone surface to implant contact and a more rapid healing response when compared to pure titanium fixtures (Weinlander et al., 1992; Gottlander et al., 1992). After implantation, hydroxylapatite coated fixtures develop an average of five to eight times greater mean interfacial strength than uncoated, grit blasted titanium implants. (Cook et al., 1987). Although many coated implants are implanted it is not yet clear whether or not long-term benefits will result.

Calcium phosphate compounds such as hydroxylapatite, oxyhydroxylapatite, and tricalcium phosphate stimulate osteophilic or osteoconductive biologic activity in the absence of a foreign body response. (Jarcho et al., 1977; de Groot, 1980; Bagambisa et al., 1990). At the electron microscopic level hydroxylapatite implants appear to form a tight interface with the bone that is described as biointegration (Meffert et al., 1987). Bone matrix is proposed to be deposited on the hydroxylapatite layer due to physico-chemical interaction between collagen of the bone and hydroxylapatite crystals of the implant (Denissen et al., 1986). Tissue compatibility is excellent. However, the fragility of calcium phosphate compounds prevent them from being used as structural components.

If superior tissue response can be achieved with a stable composite of calcium phosphate and a strong inert material that does not weaken, tooth root and other structural applications may be possible (McGee, 1974). A two-phase mixture of calcium phosphate and magnesium aluminate spinel, a biologically active ceramic/ceramic composite (osteoceramic), was first studied by Janikowski and McGee in 1969. In this combination, the calcium phosphate enhances tissue response and the spinel phase provides mechanical strength and longevity (McGee et al., 1995a or 1995b). Spinel is insoluble in aqueous solutions and is much stronger than the calcium phosphates (McGee and Wood, 1974).

The 18 month dog study described below characterizes and compares bone and soft tissue responses of osteoceramic composite implants (Tweden, 1987) to that of pyrolytic carbon, single-crystal sapphire, and titanium-aluminum-vanadium alloy root form endosseous dental implants.

MATERIALS AND METHODS

Pyrolytic carbon (Pyrolite®, Calcitek, Inc., San Diego, CA), single crystal sapphire (Bioceram®, Kyocera, Japan) and titanium alloy (6%Al, 4%V, Core-Vent, Core-Vent Corp., Encino, CA) implants were provided sterile and used as received (Figure 1). The osteoceramic implants were fabricated and sterilized using steam as described elsewhere (Tweden, 1987).

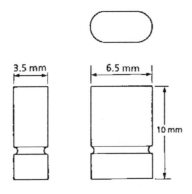

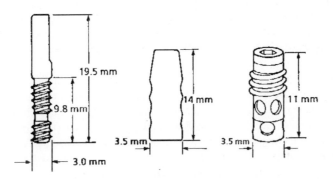

Figure 1.   Geometry of the implants: Above, osteoceramic
               Below: Sapphire, pyrolytic carbon and titanium alloy, left to right

The physical properties of the implants are quite different (Table 1).

Table 1 Physical Properties

	Density	Compressive Strength	Tensile Strength	Young's Modulus
	(gm/cc)	(MPa)	(MPa)	(GPa)
Pyrolytic Carbon	1.7-2.2	1200	69-207	0.17-0.28
Sapphire	3.97	2940	1280	392
Titanium Alloy	4.50	860	860	110
Osteoceramic	2.97	299	70	114

Implantation

Carbon, single crystal sapphire, titanium alloy (6%A1, 4%V) and osteoceramic composite (McGee, 1974) implants were inserted into healed first, second, and third mandibular premolar extraction sites in 10 adult, mixed-breed dogs. Each dog received all four implants using surgical techniques recommended by the respective manufacturers.

Dogs were placed under anesthesia, mucoperiosteal soft tissue flaps were raised and implant sites were prepared using recommended slow speed drills (700-800 rpm) with copious saline irrigation. Internal irrigation was used for drilling the titanium alloy implants sites while external irrigation was used for the other three implants. Site location was chosen at random.

The pyrolytic carbon (3.5 mm O.D. X 7 mm L) and the sapphire implants (3 mm O.D. x 19.5 mm L) penetrated the oral mucosa immediately following insertion (transgingival, single stage design).

The titanium alloy (3.5 mm O.D. x 11 mm L) and osteoceramic (3.5 mm O.D. x 6.5 mm L) implants are of a two-stage design. The device is first implanted below the gingiva for wound healing after which a second surgery is done to attach an abutment. The implant is placed in function through the connection of a crown or prosthodontic appliance. Second stage surgery was not done in this study.

NIH guidelines for the care and use of laboratory animals (NIH Publication #85-23 Rev. 1985) were followed. The dogs were fed a soft diet for six weeks after implantation and then returned to a hard diet. Pre and post-implantation radiographs were taken and the dogs were given an antibiotic for five days post-implantation (Tribrissen, Burroughs Wellcome Co., Research Triangle Park, N.C.). To assess bone healing fluorescent labeling was achieved by giving 250 mg of oxytetracycline and 300 mg of demeclocycline at 8 hour intervals for three days. Oxytetracycline was started 21 days and demeclocycline started 10 days prior to animal sacrifice.

Clinical Evaluation

At 1, 2, 3, 6, 9, 12 and 18 months clinical parameters were measured and photographs and radiographs were taken of the implant sites. Gingival health (absence of redness and bleeding), plaque accumulation, gingival sulcus (probing) depth and degree of mobility were evaluated by an experienced periodontist (GIM) and radiolucency surrounding the implants was measured by a veterinary orthopedic surgeon (CLR). Clinical and radiographic rating scales (McKinney et al. 1982) began at zero and increased to 4 as the parameter being evaluated deteriorated. The right and left fourth mandibular premolar served as control sites. Clinical data were analyzed statistically using the analysis of variance method for each period.

Histologic Evaluation

Two dogs each were sacrificed for histologic examination at 3, 6, 9, 12 and 18 months. Block sections of the mandibles containing the implants were fixed in 70% ethanol and dehydrated by 24 hour exposures to 95% and 100% (X2) ethanol. After embedding in epoxy (Spurr's low viscosity embedding media, Polysciences, Inc., Warrington, PA), a low speed sectioning saw (Isomet low speed saw, Buehler, Ltd., Lake Bluff, IL) was used to cut facial to lingual 300 micron thick sections. These sections were glued to glass slides, ground to approximately 70-100 micron thickness and stained with "Paragon" multiple stain (Paragon C and C Co., Inc., Bronx, NY). Unmounted sections were microradiographed with a Torrex X-ray Inspection System operating at 60 Kv for 10 seconds.

Labeled bone growth was evaluated using an Olympus AS-2 Fluorescent microscope and analyzed using blue Epi illumination on sections stained 10 µm deep with Paragon.

RESULTS

Clinical Evaluation

Probing depth values displayed a significant difference in means between the control teeth (3 mm) and both sapphire (3.4 mm) and pyrolytic carbon (3.9 mm) implants at one month post-implantation, $p = 0.984$ and $p = 0.999$ respectively. A general upward trend in plaque, bleeding, mobility and radiographic index values were observed with time, suggesting adverse tissue reactions to the sapphire and pyrolytic carbon implants. One sapphire implant was lost 16 months post-implantation. Radiographic assessment was not possible for pyrolytic carbon due its high radiolucency. Only three pyrolytic carbon implants remained at three months making comparisons difficult. Nine of the ten carbon implants failed before six months.

Titanium alloy and osteoceramic implants were placed below the mucosa, therefore gingival inflammation, plaque accumulation and mobility indices were not measured for these implants. However, inflammation and fistulas occurred over some titanium implants up to twelve months after surgery. At one month post-implantation, radiolucency index around the titanium alloy implants (0.9 out of 4) was greater than the control teeth (0) and the osteoceramic implants (0.3). Radiolucency for the titanium alloy implants peaked at 6 months after which it decreased. Radiolucency remained low for the osteoceramic implants for the full 18 months of the study.

Histologic Results Using Transmitted Light and Microradiography

Percent of the implant surface area in contact with bone was determined using light microscopic examination. Bone contact values were averaged at each time period and plotted for each implant (Figure 2).

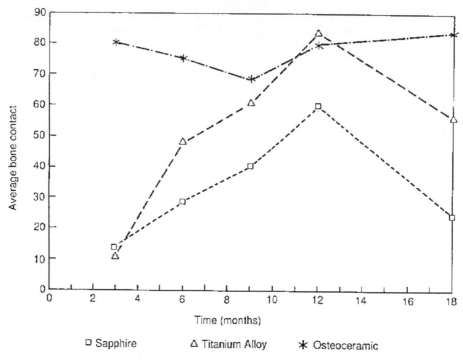

Figure 2.   Average bone contact (percent versus time)

Only one carbon implant could be evaluated histologically because of the high failure rate. This implant was retrieved at 12 months post-implantation. Despite the apparent lack of mobility of this implant, there was no bone adaptation to the carbon in any of the sections. Clusters of inflammatory cells were present at the epithelial-implant interface and a downward migration of the surface epithelium was observed.

In general, bone in contact with the sapphire implants was relatively low for the first nine months including two sapphire implants which were completely encapsulated with fibrous tissue. Good repair and adaptation of bone was observed adjacent to one sapphire implant at 12 months (Figure 3); however, the epithelial-implant interface contained degenerated tissue and a moderate inflammatory reaction. In addition, one sapphire implant could be rotated and depressed at 12 months and was lost at 16 months. Seventy-eight percent of the sapphire implants were associated with various degrees of fibrous tissue migration.

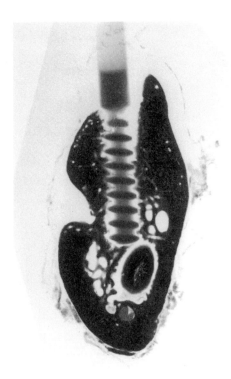

Figure 3.   Sapphire implant 12 months (x12)

Bone contact to titanium alloy implants was low at three months (Figure 2) and gradually increased up to 12 months post-implantation. A moderate inflammatory reaction and disorganized fibrous connective tissue was noted with these implants throughout the study. Bone grew into the vents of most of the implants by 6 months post-implantation (Figure 4,5). Inflammatory cells were found associated with the polysulfone inserts which were placed into the top half of the implants during the healing stage of the bone. At 12 months, 83% bone contact was observed with the titanium alloy implants which decreased to 56% by 18 months (Figure 2). Well-formed Haversian systems were observed in contact with most of the implants at 12 and 18 months.

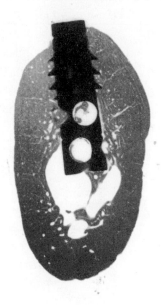

Figure 4.   Microradiograph of a titanium alloy implant at nine months (x3)

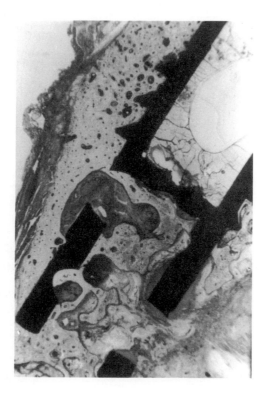

Figure 5.   Transmitted light section of a titanium alloy at nine months (x12)

The osteoceramic implant showed high bone contact (80%) at 3 months and maintained a high level of contact throughout the study (Figure 2). In some cases, a significant amount of bone growth was observed covering the superior aspect of the osteoceramic implant that had been placed level with the top of the alveolar ridge. Well-formed Haversian systems were observed adjacent to the osteoceramic implants. Microradiographs showed dense compact bone in contact with the osteoceramic implants in all cases (Figure 6). In several cases, cracks in the osteoceramic implant produced during sectioning were found to continue into the bone. This phenomenon indicates a strong bond existed between the implant and the bone.

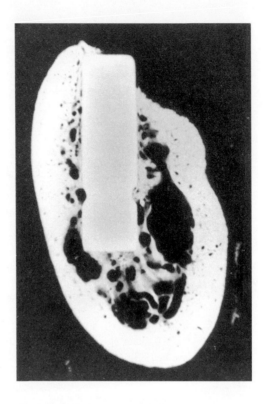

Figure 6.   Microradiograph of an osteoceramic at three months (x3)

Histologic Results Using Fluorescent Ilumination

Epi illumination and fluorescent microscopy allowed for evaluation of healing responses as related to time. Sections from each time period for sapphire, titanium alloy and osteoceramic implants were examined around their entire periphery. Areas of calcium deposition at the time the labels were administered could be compared between implants and at various locations around existing implants. The Paragon stain gave contrast to the soft tissue and fluorescence gave contrast to the calcium deposition sites. Sapphire and titanium implants were found to be embedded in dense bone at periods up to nine months with fluorescent microscopy (Figure 7,8). However, at 12 and 18 months, the sapphire implants were almost completely surrounded by a thin fibrous capsule (Figure 9). At 18 months, the titanium implants had large areas of fibrous tissue invasion (Figure 10). Fluorescent microscopy confirmed that osteoceramic implants were fully incorporated in bone at three months and for the duration of the study (Figure 11,12). Fluorescent osteonal canals were observed around implants with bone contact in histologic sections.

Figure 7.   Fluorescent micrograph of a sapphire at nine months showing good bone contact and remodeling (x45)

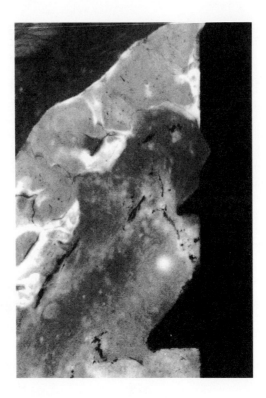

Figure 8.   Fluorescent micrograph of titanium implant at 3 months showing good bone registry and no remodeling

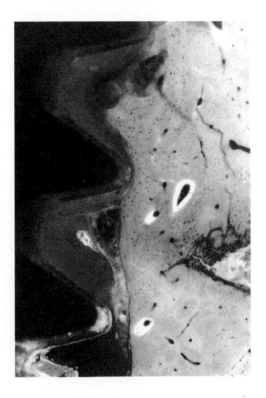

Figure 9.   Fluorescent image of a sapphire implant at 18 months (x45)

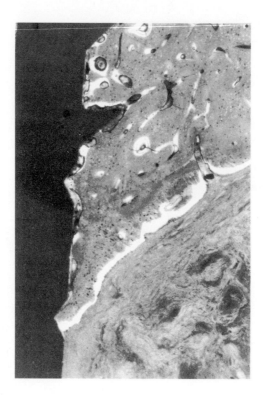

Figure 10.   Fluorescent image of a titanium implant at 18 months (x45)

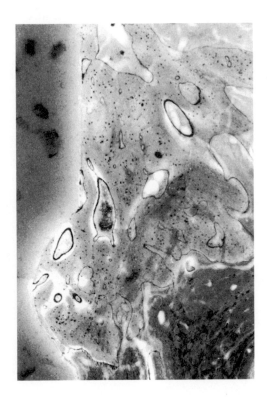

Figure 11.   Fluorescent image of osteoceramic notch section at 3 months (x45)

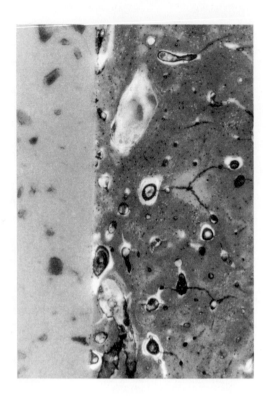

Figure 12.   Fluorescent image of osteoceramic at 18 months (x45)

DISCUSSION

This study characterized and compared a ceramic/ceramic composite (osteoceramic) two-stage implant with that of three other implant designs and materials: Pyrolytic carbon and single crystal aluminum oxide (sapphire) implants designed for a single stage transgingival surgical placement and a titanium-aluminum-vanadium alloy root form implant. Both the osteoceramic and the titanium alloy implants are designed to be surgically placed submucosal (stage one) followed by a second surgical procedure to expose the implant for loading (stage two). Because this study was designed to assess healing responses to implant materials stage two surgery was not completed, and implant loading was not analyzed.

Osteoceramic and titanium alloy implants showed greater bone contact compared to the transgingival implants. Superior bone contact of the osteoceramic at 3, 6, 9 and 18 months indicates that it has physico-chemical properties of a biologically-active composite (Denissen et al. 1986) allowing for more rapid and complete osseointegration. Gottlander and Albrektsson (1992) demonstrated at six months that bone contact to a hydroxylapatite coated implant was higher (range 63.9-87.6%) than that seen with a similarly designed pure titanium implant (range 51.3-71.5%). However, at one year post-implantation a significant reversal occurred where bone contact with the coated titanium implant was characterized by fragments of hydroxylapatite separating from the implant surface (Carlsson et al. 1994). In the present study the osteoceramic showed greater bone contact earlier (80%), and the ability to maintain this high level of contact, which suggests a superior biologic response over hydroxylapatite coated titanium implants.

There is no correlation of the physical properties (Table 1) with the tissue contact areas and the tissue response. The stiffness of the titanium alloy and the osteoceramic are very similiar but the osteoceramic had superior tissue response. The high stiffness of the sapphire and the bone-like stiffness of the pyrolytic carbon are extremely different. But the sapphire response was superior to the pyrolytic carbon. The chemical activity of the osteoceramic appears to be more important than the physical properties in controlling tissue response.

Single stage titanium implants were placed in a protected submucosal environment using a procedure that allows the self-tapping threads of the implant to anchor it firmly in dense bone. However, after 18 months large portions of the threads were encapsulated with fibrous tissue.

Sapphire implants were rigidly fixed at placement by engaging the implant threads into a previously tapped bone site. However, one specimen became depressible and rotatable and was lost at 16 months. Fibrous tissue encapsulation increased over time with a greater area of fibrous tissue in the apical portion of the implant. There was no tissue inflammation noted in the alveolar bone adjacent to the apical fibrous tissue suggesting that encapsulation was not the result of an independent inflammatory response. Sulcus depth increased and gingival health deteriorated over time.

Only one pyrolytic cabon implant survived to the scheduled retrieval time (12 months). After wound healing is complete, physical stresses are believed to stimulate bone growth. However, motion before wound healing is complete can prevent bone attachment. Clearly the carbon implant design did not provide for adequate early stabilization since typical external forces dislodged most of the implants. Because the sapphire implants are threaded into bone, excessive motion should not be possible. This lack of clinically detectable mobility generally proved to be true with the sapphire implants. However, clinically undetectable micromovements caused by external forces on the implants may have led to the fibrous tissue encapsulation seen with this implant and the single stage implants in general (Brunski, 1988).

Transgingival implants also are subject to infection from oral cavity microorganisms. The success of transgingival implants depends strongly on oral hygiene to prevent infection and concomitant bone loss. Antibiotics administered post surgery and prior to sacrifice were ineffective in eliminating microbiol plaque formation and gingival inflamation. It is possible infection contributed to the loss of the single stage implants. The presence of an inflammatory response to the sapphire implants was observed in the sections. Inflammation and bone resorption was observed for the sapphire implants in the margins nearest the gingiva, especially at the longer evaluation periods. Clusters of inflammatory cells were also seen at the epithelial-implant interface of the carbon implant at 12 months.

With strength comparable to compact cortical bone and a modulus of elasticity of enamel (Tweden, 1987), osteoceramic could serve as an effective biomechanical material for dental and medical applications. Recent technology using a plasma spray of hydroxylapatite to coat titanium results in amorphous calcium phosphate layer that lacks the bioactivity of crystalline hydroxylapatite. Dennissen et al. (1986) reported plasma sprayed coatings are susceptable to dissolution, possibly due to loss of crystallinity during the high temperature plasma spraying procedure. In addition, bacteria have been shown to digest or dissolve plasma sprayed hydroxylapatite (Verheyen et al. 1993). Dissolution of hydroxylapatite coatings have been proposed as a reason for failure of HA coated implants (Johnson, 1992). The hydroxylapatite-titanium interface has been reported to be weaker than the interface formed between hydroxylapatite and bone (Block et al. 1990). The osteoceramic is a homogenous ceramic without a metal-calcium phosphate interface found in coated titanium implants. In addition, osteoceramic appears to have bioactivity comparable to hydroxylapatite and can be processed to fit specific human anatomical applications.

Separation of material and design factors is difficult when discussing reasons for the success or failure of the implants in this study; however, significant differences were seen among the implants. Two-stage implants (osteoceramic and titanium alloy) had greater bone contact than single-stage implants that protruded into the oral cavity (sapphire and pryrolytic carbon). This observation supports the advantage of two stage implant systems which allow for osseointegration in a closed environment before loading with functional forces. Osteoceramic was found to be a biologically active ceramic-ceramic composite with bone contact superior to the other materials throughout the studies. Osteoceramic was not walled off by a fibrous capsule while the other implants were. Because osteoceramic appears to have excellent tissue compatibility, significant bony integration and adequate strength, further research is necessary to evaluate this ceramic/ceramic material under load and to compare it with implants currently in clinical practice.

ACKNOWLEDGEMENT

We thank Dr. Sandra McNeal for evaluating the clinical radiographs; the Iowa State University Research Foundation for financial support; and Dr. Christopher Squier, Dental College, University of Iowa, for providing facilities for microradiography.

REFERENCES

1)   Albrektsson, T., Dahl, E., Enborn, L., Engevall, S., Engquist, B., Erikksson, A.R., Feldman, G.,
      Freibert, N., Glanta, P., Kjellman, O., Kristersson, L., Kvint, S. (1988): Osseointegrated Oral
      Implants-- A Swedish Multicenter Study of 8139 Consecutively Inserted Nobelpharma Implants, J.
      Periodontol., 59: 287-296.

2)   Bagambisa, F.D., Joos, U., and Schilli, W. (1990): The interaction of Osteogenic Cells With
      Hydroxylapatite Implant Materials In Vitro and In Vivo, Int. J. Oral Maxillofac. Implants, 5, 217-
      226.

3)   Block, M.S., Delgado, A., Fontenot, M.G. (1990): The Effect of Diameter and Length of
      Hydroxylapatite-coated Dental Implants on Ultimate Pullout Force in Dog Alveolar Bone. J. Oral
      Maxillofac. Surg. 48, 174-178.

4)   Brunski, J. (1988): Biomaterials and Biomechanics, CDAJ: 66-77.

5)   Carlsson, L., Regner, L., Johansson, C., Gottlander, M., and Herberts, P. (1994): Bone response to
      hydroxyapatite-coated and commercially pure titanium implants in the human arthritic knee. J.
      Orthop. Res., 12(2): 274-285.

6)   Cook, S.D., Kay, J.F., Thomas, K.A., and Jarcho, M. (1987): A Histomorphometric Study of
      Unthreaded Hydroxylapatite-coated and Titanium-coated Implants in Rabbit Bone, Int. J. Oral
      Maxillofac. Implants, 7,485-490.

7)   DeGroot, K. (1980): Bioceramics Consisting of Calcium Phosphate Salts, Biomaterials 1, 47-50.

8)   Denissen, H.W., Veldhuis, A.A., and van der Hooff, A. (1986): Hydroxylapatite Titanium
      Implants. Proceedings of International Congress on Tissue Integration In Oral and Maxillofacial
      Reconstruction. May 1985. Brussels. Excerpta Medica, Current Practice Series #29, 372-389.

9)   Gottlander, M., and Albrektsson, T. (1992): Histomorphometric Analysis of Hydroxylapatite-
      coated and Uncoated Titanium Implants. The Importance of Implant Design. Clin. Oral Impl. Res.
      3, 71-76.

10)   Gottlander, M., Albrektsson, T., and Carolsson L.V. (1992): A Histomorphometric Study of
      Unthreaded Hydroxylapatite-coated and Titanium-coated Implants in Rabbit Bone, Int. J. Oral
      Maxillofac. Implants, 7: 485-490.

11)   Janikowski, T. and McGee, T.D. (1969): Artificial Teeth for Permanent Implantation, Proc. Ia.
      Acad. Sc. 76, 113-118.

12)   Jarcho, M., Kay, J.F., Guimaer, K.I., Doremus, R.H., and Drobeck, H.P. (1977): Tissue, Cellular
      and Subcellular Events at a Bone-Ceramic Hydroxyapatite Interface, J. Bioengr. 1: 70-92.

13) Johnson, B. (1992): HA-coated Dental Implants: Long-term Consequences. J. Calif. Dent. Assoc. 20: 33-41.

14) McGee, T.D. (1974), US Patent 3,787,900.

15) McGee T.D., Graves, A.W., Tweden, K.S. and Niederauer, G. (1995a): "A Biologically Active Ceramic Composite with Enduring Strength" Part A, Vol. 2, Chapt 42. Encyclopedic Handbook of Biomaterials and Bioengineering. p.p. 1413-1427.

16) McGee, T.D., and Olson, C.E. (1995b): "General Requirements for a Successful Orthopedic Implant", Part B, Vol. 1, Chapt 3. Encyclopedic Handbook of Biomaterials and Bioengineering. p.p. 69-82.

17) McGee, T.D. and Wood, J.L. (1974): Calcium-Phosphate Magnesium Aluminate Osteoceramics, J. Biomat. Res. 5: 137-144.

18) McKinney, R.J. Jr., Koth, D.L. and Steflick, D.E. (1982): The Single Crystal Sapphire Endosseous Dental Implant. II Two Year Results of Oral Trials, J. Oral Implant. 10: 619-638.

19) Meffert, R.M. (1985), Iowa Society of Periodontology: Mar 1.

20) Meffert, R.M., Block, M.S., and Kent, J.M. (1987): What is Osseointegration? Int J. Periodontics Restorative Dent., 4, 9-14.

21) Schnitman, P.A., (1993): Implant Dentistry: Where Are We Now?, JADA, 124: 39-47.

22) Stillman, N. and Douglas, C.W. (1993): The Developing Market For Dental Implants, JADA, 124: 51-56.

23) Tweden, K.S. (1987), A Comparison of Four Endosseous Dental Implants: Single Crystal Sapphire; Pyrolitic Carbon; an Alloy of Titanium, Aluminum and Vanadium; and a Biologically Active Ceramic Composite Consisting of Calcium Phosphate and Magnesium Aluminate Spinel, Ph.D. Dissertation, Iowa State University.

24) Verheyen, C., deWijn, J., Van-Blitterswijk, C., deGroot, K., and Rozing, P. (1993): Hydroxlapatite/poly (L-lactide) Composites: an Animal Study on Push-out Strengths and Interface Histology, J. Biomed. Mater. Res., 27: 433-444.

25) Weinlander, M., Kenney, E.B., Lekovic, V., Moxy, P., Lewis, S., (1992): Histomorphometry of Bone Around Three Types of Endosseous Dental Implants, Int. J. Oral Maxillofac. Implants, 7: 491-496

Materials Science Forum Vol. 293 (1999) pp. 37-64
© 1999 Trans Tech Publications, Switzerland

Bioactive Glasses and Glass-Ceramics

L.L. Hench[1,2]

[1] Department of Materials Science and Engineering, University of Florida, Gainesville, FL, USA

[2] now at: Department of Materials, Imperial College of Science and Technology, London, UK

Keywords: Bioactive, Bioglass®, Glass-Ceramic, Glass

ABSTRACT

In 1969, certain compositions of glasses and glass-ceramics were discovered to form a mechanically strong bond with bone. The clinical application of Bioglass® and related materials grew in three different directions through the 1970's (as bioactive glasses, synthetic hydroxyapatite, and bioactive glass-ceramics). Specific examples include bioactive coatings, non load-bearing implants, load-bearing implants, bioactive particulates, bioactive cement, and bioactive composites. Future applications of these materials will be enhanced by understanding the genetic basis for cell proliferation. It may be possible to design bioactive compositions conducive to the repair of osteoporotic bone and to enhance the quality of bone bonded to the bioactive implant.

THE BEGINNING

In 1969, it was first discovered that certain compositions of glasses and partially crystallized glass-ceramics in the $Na_2O-CaO-P_2O_5-SiO_2$ system (Table 1) formed a mechanically strong bond with living bone. [1] This behavior, now termed bioactive bonding, [2-6] was in marked contrast with other implant materials (such as stainless steel, polymethylmethacrylate (PMMA), or alumina) which elicit formation of a thin, fibrous capsule that isolates the material from bone. X-ray diffraction analyses showed that a crystalline hydroxyapatite (HA) layer developed on the glass surface when it was exposed *in vivo* and to simulated body fluids (SBF). [1] Transmission electron microscopy (TEM) of the bonded bone-implant interface at 6 weeks showed that normal bone cells (osteoblasts) were present. [1] The osteoblasts were producing collagen fibers which became incorporated within the growing HA layer. *In vitro* experiments in solutions containing calcium and phosphate ions showed that the HA layer formed on the implant was due to heterogeneous nucleation and growth of crystalline HA using ions both from the glass and the solution. [1]

The mechanical strength of the bone-implant interface was sufficiently high that either the bone or glass broke first when load was applied. [1] The interface did not fail. Even diamond microtome preparation of TEM thin sections failed to dislodge bone from the implant bonding zone. The TEM analysis showed intimate interdigitation of HA crystals and collagen fibers at an ultrastructural, nanometer, level. [1]

Based upon an analysis of alternative types of chemical bonding at interfaces, the Hench et al. paper proposed that the strong mechanical adherence of the bioactive glass and bone was due to carboxyl bonds of collagen fibril end groups with calcium and phosphate sites on HA crystals. [1] Such bonding is equivalent to the inorganic-organic bonds of normal bone. [7] Partial crystallization of the glass prior to implantation had little effect on the interfacial bonding and therefore the conclusion was reached that the amorphous glassy phase was important in forming the bond. [1]

Subsequent papers showed that the bioactive material, termed Bioglass®, released soluble Si, Ca, and P ions into solution very rapidly due to both ion exchange with H^+ and H_3O^+ and by silicate network dissolution. [8,9] An important effect of ion exchange was the rapid shift of the interface towards alkalinity which accelerated the heterogeneous precipitation of hydroxyapatite crystals and binding of collagen fibrils. Auger electron spectroscopy profiles of the bioactive glass surface showed a Ca, P-rich layer growing on a SiO_2-rich layer within hours after implantation. [9] Collagen was bound in the growing layers.

GROWTH OF THE FIELD

By the mid 1970's, the field of bioactive implants had begun to grow in three different directions. The studies at the University of Florida continued with emphasis on bioactive glasses. [10-14] Their efforts were devoted towards understanding the interfacial bonding mechanisms and potential orthopaedic applications. The second growth direction, headed by Michael Jarcho and colleagues in the US, Klaas de Groot and co-workers in Europe, and H. Aoki's team in Japan, was use of synthetic hydroxyapatite implants, primarily in dental applications. [15-19] The rationale was straightforward; i.e., since the mechanism of bonding of bioactive glasses is by means of forming an HA layer *in situ*, then a synthetic HA implant should also develop a bond with bone. Kay showed that the rationale was correct and bone did bond to HA implants at an ultrastructural level, as revealed by TEM analysis of bonded interfaces. [20] Clinical applications soon followed. [15-19]

The third growth direction was to use multi-phased glass-ceramics to improve the mechanical properties of the bioactive glasses. The efforts of Broemer and Deutcher in Germany, [6] leading to Ceravital® (Table 1), confirmed the findings of Hench, et al. [1] The bioactive glass-ceramic implants produced a mechanically strong bond with bone and led to extensive new bone growth. However, Gross et al. showed that there was a problem of *in vivo* interphase attack of the multiphase material which limited clinical applications. [6] Gross's team also established the limitations of compositional change of bioactive implants by showing that small additions of multivalent ions (Al^{3+}, Ta^{5+}, etc.) prevented bone bonding.

The Japanese effort in this direction, led by T. Yamamuro and T. Kokubo's groups at Kyoto University, was more fruitful. [4, 5, 21-23] They developed a hot pressed, fine grained bioactive glass-ceramic composed of apatite and wollastonite ($CaO\bullet SiO_2$) with a small amount of residual glassy phase (Table 1). The A/W glass-ceramic (Cerabone®) has a very uniform microstructure with small grain size and is resistant to interphase boundary attack. The mechanical properties of A/W glass-ceramic are excellent (Table 1) and it is now used clinically, especially in vertebral reconstruction and iliac crest repair. [4]

Table 1. Composition and Mechanical Properties of Bioactive Ceramics Used Clinically

Ref.	1	2	3	4	5	6	7-9	10, 11
	Bioglass® 45S5	S53P4	Glass-Ceramic Ceravital®	Glass-Ceramic Cerabone® A-W	Glass-Ceramic Ilmaplant® L1	Glass-Ceramic Bioverit®	Sintered HA* >99.2%	Sintered β-TCP* >99.7%
Composition (wt%)								
Na_2O	24.5	22.6	5-10	0	4.6	3-8		
K_2O	0		0.5-3.0	0	0.2	3-8		
MgO	0		2.5-5.0	4.6	2.8	2-21		
CaO	24.5	21.8	30-35	44.7	31.9	10-34		
Al_2O_3	0		0	0	0	8-15		
SiO_2	45.0	53.9	40-50	34.0	44.3	19-54		
P_2O_5	6.0	1.7	10-50	16.2	11.2	2-10		
CaF_2	0			0.5	5.0	3-23		
B_2O_3	0							
Phase*	Glass	Glass	Apatite Glass	Apatite β-Wollastonite Glass	Apatite β-Wollastonite Glass	Apatite Phlogopite Glass	HA	Whit-lockite
Density (g/cm^3)	2.6572			3.07		2.8	3.16	3.07
Hardness Vickers (HV)	458 ± 9.4			680		500	600	
Compressive Strength (MPa)			500	1080		500	500-1000	460-687
Bending Strength (MPa)	42(Tensile)			215	160	100-160	115-200	140-154
Young's Modulus (GPa)	35		100-150	118		70-88	80-110	33-90
Fracture Toughness K_{Ic} $(MPa \bullet m^{1/2})$				2.0	2.5	0.5-1.0	1.0	
Slow Crack Growth (n)				33			12-27	

* HA = $Ca_{10}(PO_4)_6(OH)_2$, βTCP = Whitlockite = $β-3CaO \bullet P_2O_5$, Apatite = $Ca_{10}(PO_4)_6(OH,F)_2$, β-Wollastonite = $CaO \bullet SiO_2$, Phlogopite = $((Na,K)Mg_3(AlSi_3O_{10})F_2)$

1. Hench, L.L. and Ethridge, E.C.: p. 137 in *Biomaterials, An Interfacial Approach*, 1982, Academic Press, New York.
2. Andersson, O.H., et al.: *Glastech.Ber.*, 1988, **61**, 300.
3. Bromer, H., et al.: p. 219 in *Science of Ceramics*, Vol. 9, 1977.
4. Kokubo, T.: in *Multiphase Biomedical Materials*, 1989, VSP, Utrecht.
5. Berger, G., et al.: p. 120 in *Proc. XV Intl. Cong. Glass*, Vol. 3a, 1989, Nauka, Leningrad.
6. Vogel, W. and Holland, W.: *Angew Chem.Int.Ed.Engl.*, 1987, **26**, 527.
7. Jarcho, M, et al.: *J.Mater.Sci.*, 1976, **11**, 2027.   8. Akao, M, et al.: *J.Mater.Sci.*, 1981, **16**, 809.
9. Dewith, G., et al.: *J.Mater.Sci.*, 1981, **16**, 1592. 10. Jarcho, M, et al.: *J.Mater.Sci.*, 1979, **14**, 142.
11. Akao, M., et al.: *J.Mater.Sci.*, 1982, **17**, 343.

THE PRESENT: LIMITATIONS VERSUS APPLICATIONS

Clinical uses of bioactive implants from all three growth directions (bioactive glasses, synthetic HA, and bioactive glass-ceramics) have revealed limitations of this class of biomaterials. No single material is best for all clinical needs. Table 2 shows that a bioactive material can be used in various forms: bulk, coatings, powders, or composites. It is essential that the material in a specific form be matched to the clinical function required. Examples of efforts to optimize the material, form and clinical function follow.

Table 2. Form, Phase and Function of Bioactive Implants

Form	Phase	Function
Powder	Polycrystalline Glass	Space-filling, therapeutic treatment, regeneration of tissues
Coating	Polycrystalline Glass Glass-Ceramic	Tissue bonding, corrosion protection
Bulk	Polycrystalline Glass Glass-Ceramic Composite (Multi-Phase)	Tissue bonding Replacement and augmentation of tissue, replace functioning parts, space-filling

1. BIOACTIVE COATINGS

Bioactive glass and bulk HA implants do not have the strength or fatigue resistance for use as load bearing implants. An approach to minimize this limitation of strength and toughness is to use the bioactive material as a coating on a strong metal alloy used clinically, such as stainless steel or titanium alloys. HA coatings are now routinely used in orthopaedic and dental implants, although there continues to be a debate regarding the long term success of the plasma sprayed HA coatings which have dissolution rates *in vivo* that vary depending upon impurity content and extent of crystallinity. The reliability of bioactive glass coatings on metals has not been good due to interfacial attack between the metal substrate and the glass and therefore this approach to load bearing applications has been limited clinically. Chapter 9 (LeGeros and LeGeros), chapter 11 (Klein, et al.) and chapter 12 (Lacefield) in reference 3 discuss these issues in detail.

2. NON LOAD BEARING IMPLANTS

A second approach is to use a bioactive implant in a mode that minimizes mechanical loads, especially tensile stresses. Bulk Bioglass® implants have been used clinically for ten years as middle ear prostheses (figure 1) and as implants to stabilize an endosseous alveolar ridge (figure 2) (Chapter 4 in reference 3). In both applications there is very little tensile stress applied to the implant and therefore mechanical strength or fatigue resistance is not important. A Bioglass® middle ear prosthesis replaces one or more ossicles of the middle ear which have been lost due to trauma or disease, such as chronic infection (figure 2). Metallic or polymeric implants used as ossicular replacements had 3 year failure rates as high as 80% due to extrusion through the ear drum. In marked contrast, there is no evidence of extrusion of 45S5 Bioglass® middle ear prostheses after ten years of clinical trials in Florida and London. The reason for this success is bioactive bonding of the tympanic membrane (ear drum) to the implant (figure 1c).

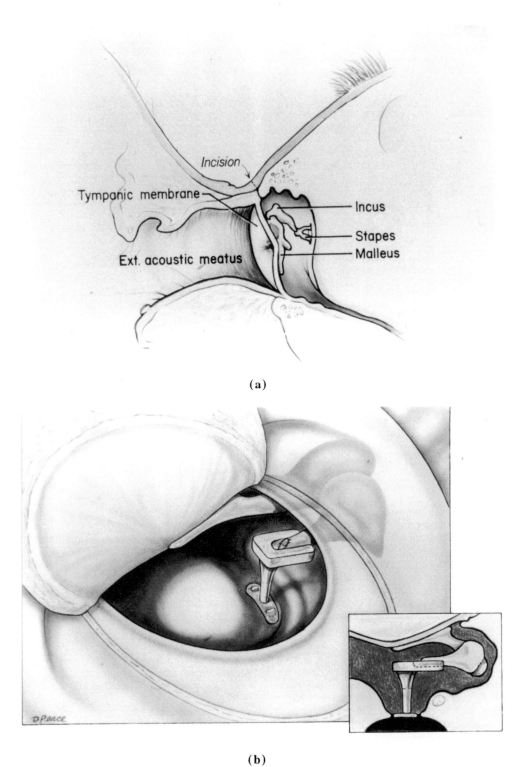

(a)

(b)

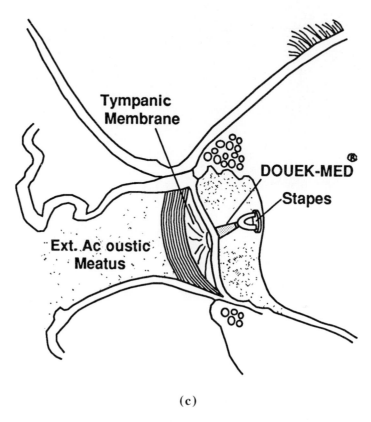

(c)

Figure 1 Bioactive glass (45S5 Bioglass®) prostheses for replacement of the ossicles of the middle ear: a) normal ear, b) partial replacement between malleus handle and stapes footplate, c) prosthesis between ear drum (tympanic membrane) and stapes footplate (DOUEK MED®).

Wilson, et al. were the first to report that only a narrow range of bioactive glass compositions (Table 1) from 42% to 52% SiO_2 formed a bond with the collagen fibers of soft connective tissues. [12] This compositional range is indicated by the cross hatched region of figure 3. Compositions that bond with soft tissues also form a bond very rapidly with bone and are designated as Class A bioactive implants. Compositions with higher percentages of SiO_2, up to a limit of 60%, form a bond only to bone and are designated as Class B bioactive implants. Glass compositions with > 60% SiO_2 are nearly inert and elicit a non adherent fibrous capsule separating the implant from bone or soft tissues (figure 3, Inert, Region B). Compositions with excessive alkali content, (figure 3, Resorbable, Region C) dissolve when used as implants. Other distinguishing features of Class A and Class B bioactivity are discussed below.

Endosseous ridge maintenance implants (ERMI®) composed of 45S5 Bioiglass® are an example of successful clinical use in a low load bearing application (figure 2). Both hard and soft tissues may occur at the ERMI interface and its alveolar bone bed and gingival tissues. There is a gradient in elastic modulus between the implant and tissue which mimics the gradient between a natural tooth, and the periodontal membrane which attaches it to bone (figure 4). Consequently, the stresses applied to the alveolar ridge by dentures are moderated by the ERMI and the bone remains healthy and does not resorb. The long term clinical success of the Class A bioactive glass implant is much greater than Class B HA implants (figure 5).

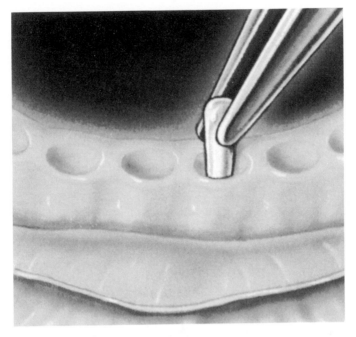

(a)

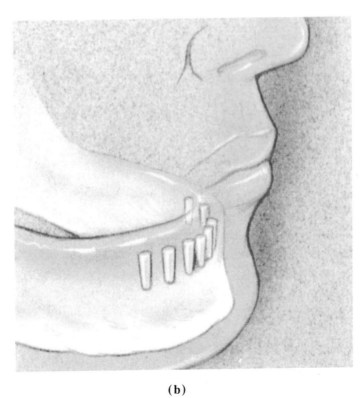

(b)

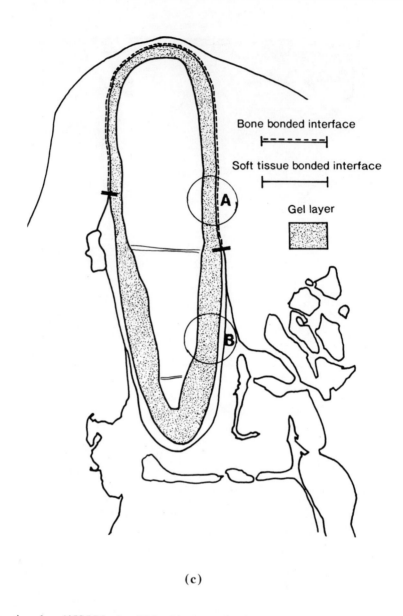

Bone bonded interface

Soft tissue bonded interface

Gel layer

(c)

Figure 2 Bioactive glass (45S5 Bioglass®) implants to maintain the alveolar bone for denture wearers: a) insertion of a Bioglass® ERMI® (endosseous ridge maintenance implant) in an extracted tooth root socket; b) ERMI's® maintaining alveolar bone; c) interfacial bonding zone between a Bioglass® ERMI® and bone. Area A is bone bonding, Area B is soft tissue bonded (based upon Wilson, et al.: *J. Oral Implantology*, 1993, **19**, 298).

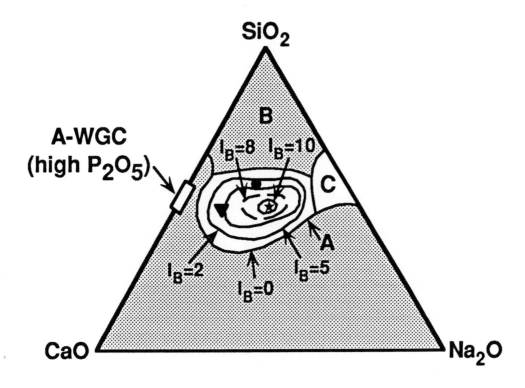

Figure 3 Compositional boundary for Na_2O-CaO-SiO_2 (with constant 6 weight % P_2O_5) bioactive glasses. Index of bioactivity (I_B) contours are shown. Region A is bone bonding, Region B is nearly inert, non bonding, Region C is resorbable. Compositions within the dashed area bond to soft connective tissues as well as bone. 45S5 Bioglass® is indicated by a star.

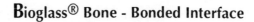

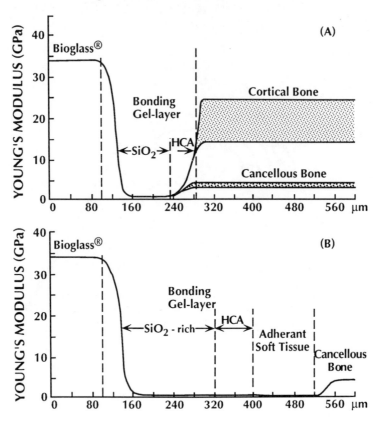

Figure 4 Elastic modulus profiles of a Bioglass® ERMI® bonded into alveolar bone as shown in figure 2; Area A is bone bonding, and Area B is soft tissue bonding. Note the thicker bonding zone between soft tissues and implant.

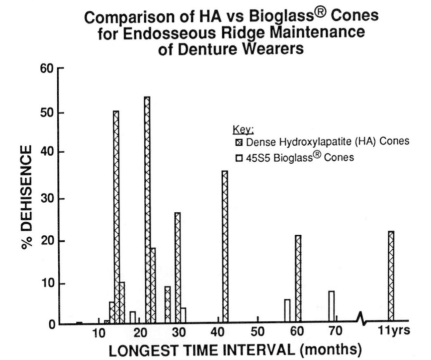

Figure 5 Comparison of Bioglass® ERMI's® showing low rates of dehiscence from the alveolar bone compared with synthetic HA implants. References for the data are given in Hench, et al.: in *Bioceramics 4*, ed. Bonfield, W.: 1990, Butterworth, Guilford.

3. LOAD BEARING IMPLANTS

The third successful approach is use of high strength bioactive A/W glass-ceramic for vertebral replacement in tumor patients. Professor Yamamuro, who pioneered this application (figure 6), reports long term success, usually the lifetime of the patient, for hundreds of cases during the last 8 years (Chapter 6 in reference 3).

Segments of autogenous bone from the iliac crest are often used in spinal and other reconstructive surgery. A/W glass-ceramic (Cerabone®) has been used widely to replace the iliac bone segment (figure 7) which shortens the recovery time for the patient (Chapter 6 in reference 3).

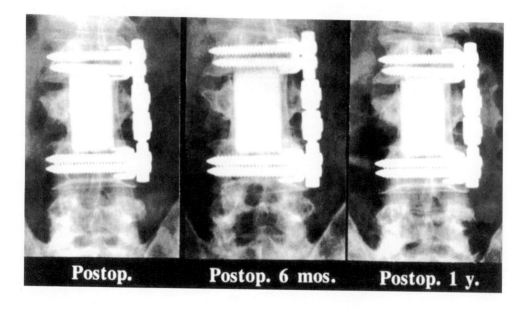

Figure 6 A/W glass-ceramic (Cerabone®) prostheses replacing vertebrae removed due to tumor. Photo from Professor Takao Yamamuro, Kyoto University.

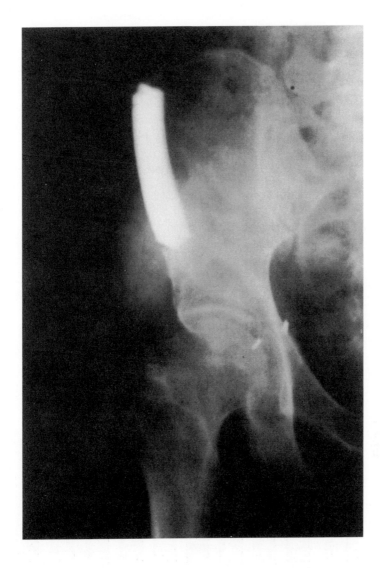

Figure 7 A/W glass-ceramic (Cerabone®) iliac crest prosthesis developed by Professor Takao Yamamuro, Kyoto University.

4. BIOACTIVE PARTICULATE

Many clinical treatments require filling in bone defects. The ideal material is the patient's own bone (autogenous bone) in the form of a paste. The growth factors in autogenous bone lead to repair of the defect within several months depending upon size and location of the defect and age of the patient. However, there are two problems with autogenous implants. The amount of material is limited and a second surgical site is often required. Freeze dried cadaver bone (heterogeneous graft) is a potential solution to these problems, but some patients may have concerns of possible viral infections from use of bone from another person. The dead bone also has to undergo remodeling before a defect is restored which delays the healing process.

Bioactive materials in the form of granules, particulate or powders can be used to fill bone defects, either alone or mixed with autogenous bone. This is a very effective means of extending a small quantity of the patient's own bone to fill a large defect. Figure 8 illustrates the effectiveness of using 45S5 Bioglass® particulate to augment the size of a canine rib. [24] A mixture of Bioglass® particulate with autogenous bone is more effective than bone alone. This study showed that Bioglass® particulates are highly effective in filling bone defects even without the use of autogenous bone.

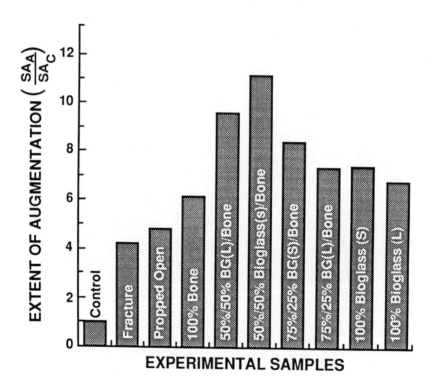

Figure 8. Effect of 45S5 Bioglass® particulate on the extent of augmentation of a dog rib bone. Large (L) and small (S) sizes of Bioglass® powders mixed with autogenous bone (BONE) are compared. Note the greater growth of bone for the Bioglass® + bone mixture than autogenous bone alone. (Based upon results reported by Wilson, et al., reference 24).

The high rate of bone formation of Class A bioactive glass powders is shown in figure 9 compared with the slower rate of bone formation using hydroxyapatite (HA) powders. Both materials, of the same particle size, were tested in the same rabbit tibial model, developed by Professor Oonishi. [25] The rapid rate of bone growth is proposed [24] to be due to the mitogenic effect of the soluble silica released by the bioactive glass which activates the genes that produce transforming growth factor-beta (TGF-β). A/W glass-ceramic particulate used alone or in combination with fibrin glue, is also more effective than HA powder in filling bone defects, as shown by Professor Yamamuro and colleagues (figure 10) (Chapter 6 in reference 3).

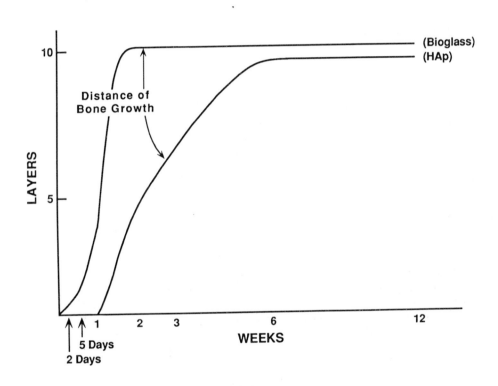

Figure 9. Comparison of bone growth in a tibial defect filled with 45S5 Bioglass® particles versus synthetic hydroxyapatite (HAp) particles. (Based upon results in Oonishi, et al., reference 25).

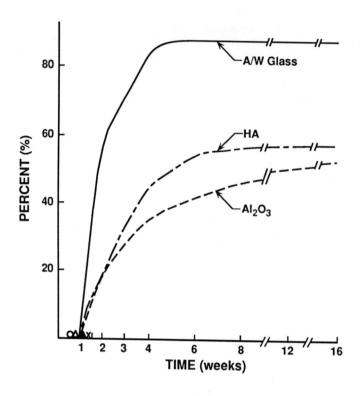

Figure 10. Comparison of bone growth in rat tibia filled with A/W glass-ceramic, HA, or Al_2O_3 based upon tests of Ono, et al.: *J.Biomed.Mater.Res.*, 1988, **22**, 869.

Particulate Bioglass® has been used to restore periodontal defects (figure 11) and other maxillofacial clinical applications for several years, with considerable success, as discussed by Wilson, et al. [26] and in Chapter 4 of reference 3. Oonishi has also used particulate HA to line the femoral canal prior to cementation of hip prostheses with good success (figure 12). The HA particulate helps shield the bone from the exothermic reaction of PMMA.

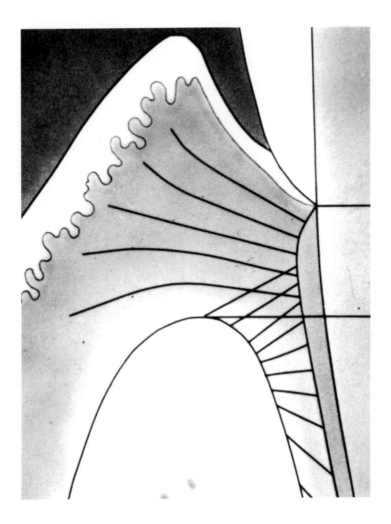

(a)

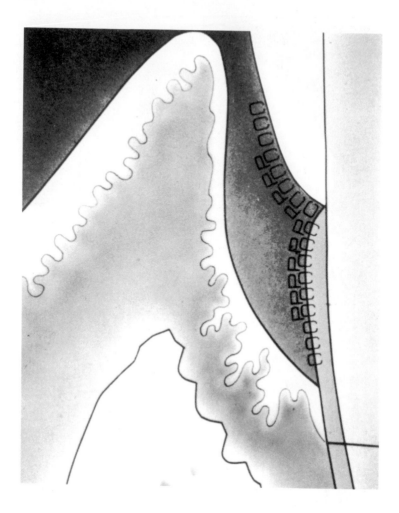

(b)

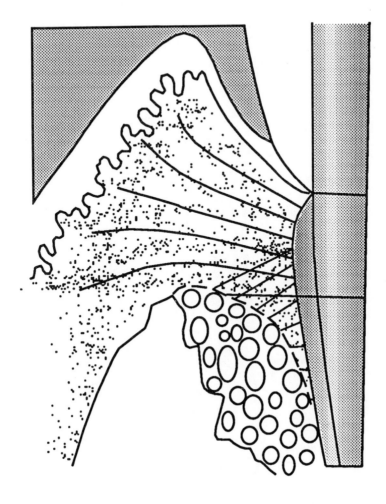

(c)

Figure 11. Effect of Perioglass® (45S5 Bioglass® particulate) on repair of bone loss due to periodontal disease. (a) Normal tooth bonded to bone by the periodontal membrane. (b) Loss of bone due to periodontal disease. (c) Restoration of bone by Perioglass®.

FEMORAL HEAD PROSTHESIS

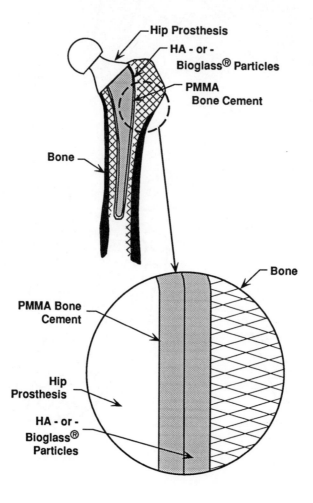

Figure 12. Use of HA or Bioglass® particles to protect femoral bone from PMMA bone cement. Based upon Oonishi, H.: p. 93 in *Bioceramics*, Vol. 6, 1993, Butterworth, Guilford.

5. BIOACTIVE CEMENT

Several forms of bioactive cement using A/W glass-ceramic granules have been developed by Professors Yamamuro and Kokubo and their teams at Kyoto University. [27] The setting times and mechanical properties are similar to polymethylmethacrylate (PMMA) bone cement. However, the bioactive cement has several advantages over PMMA; 1) an adherent bond develops between the A/W glass-ceramic particles and bone and 2) there is no large exothermic reaction during setting which prevents building of bone cells at the cement interface.

6. BIOACTIVE COMPOSITE

No implant material in use today matches the mechanical properties of bone. Strong orthopaedic metal alloys or bioceramics are very much stiffer than bone with elastic modulii 30 to 100 times larger, as illustrated in figure 13. Consequently, when such materials are fixed to bone by bone cement (morphological fixation), porous ingrowth (biological fixation) or interfacial bonding (bioactive fixation) the bone is shielded from stress and resorbs. A solution to stress shielding is to use composites with a matrix of low elastic modulus (such as polyethylene) and high strain to failure and a dispersed second phase with high elastic modulus (such as HA or Bioglass®) (Chapter 15, Ducheyne, et al. and Chapter 16 by Bonfield in reference 3 and references 28-31). A range of elastic modulii can be produced in such a composite depending upon the volume fraction of the second phase, as established by Bonfield and colleagues at the University of London, Interdisciplinary Research Center in Biomedical Materials for HA/polyethylene composites or by our University of Florida team for polysulfone/Bioglass® composites (figure 14). [13] The dispersed phase can be exposed by grinding, polishing or machining, thereby providing a bioactive interface for bonding to bone.

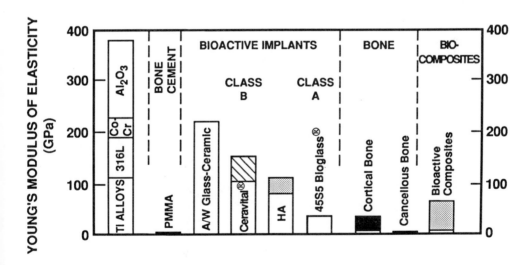

Figure 13 Comparison of Young's modulus of elasticity of non-bonding implant materials (left side), bone cement, bioactive ceramics and biocomposites with cortical and cancellous bone.

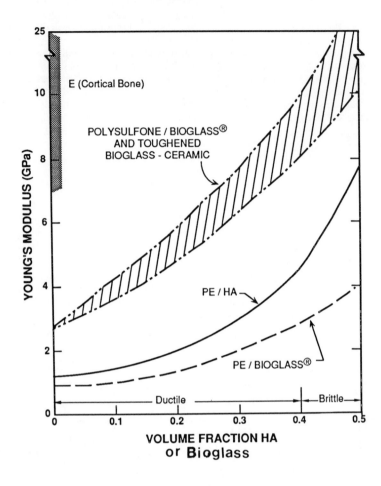

Figure 14. Range of Young's elastic modulus of bone compared with bioactive composites with varying amounts of second phase, either HA or Bioglass®.

Bone is an anisotropic material with much greater strength and strain to failure in its longitudinal direction than transverse direction (figure 15). By use of bioactive fibers as part of the second phase it should be possible to produce an anisotropic elastic modulus in a bioactive composite similar to that of bone. This approach would maximize mechanical strength of the composite where it is required in a load bearing prosthesis without raising the elastic modulus and causing stress shielding.

The uncertainty regarding composites is their reliability under cyclic fatigue conditions in the corrosive physiological environment. The interface between matrix and bioactive phase may be susceptible to crack propagation. Fatigue studies of composites under physiological loading conditions need to be performed prior to load bearing clinical applications.

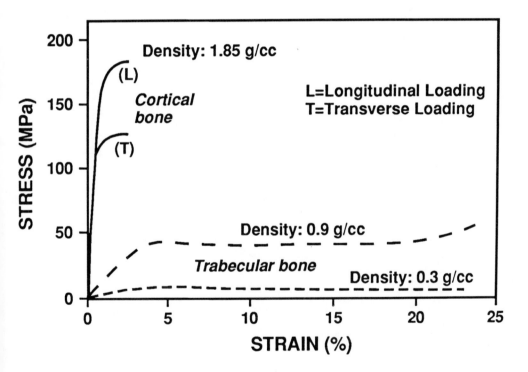

Figure 15. Comparison of stress-strain behavior of cortical bone and trabecular bone.

THE FUTURE

Biochemistry studies by Keeting, et al. [32] have demonstrated that soluble silica can activate bone cells (in culture) to produce transforming growth factor (TGF-β) which enhances cell division and bone cell proliferation. Other primary osteoblast-like cell culture experiments by Vrouwenvelder, et al. [33] have shown that bone cells grow more rapidly on Class A bioactive glass substrates than they do on Class B substrates. Thus, as summarized in figure 16, a Class A bioactive implant's surface chemistry affects cellular responses by altering the chemical environment of osteoprogenitor

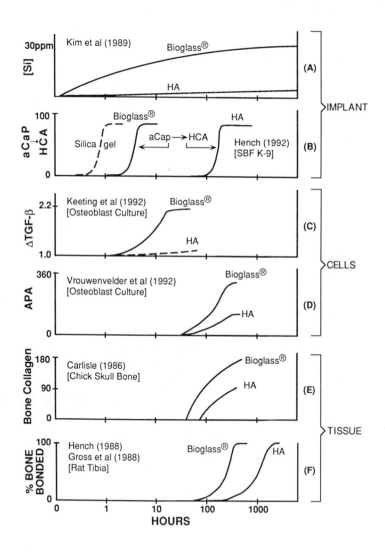

Figure 16. Series of surface reaction steps of a bioactive glass (45S5 Bioglass®) implant (Steps A and B) leading to osteoblast proliferation (Steps C and D) and rapid bone bonding and growth (Steps E and F).

stem cells. [34] The cellular response leads to an enhanced rate of bone growth. The slow dissolution of soluble Si from bioactive glasses enhances bone cell proliferation and formation of osteoid. The release of Ca and P from bioactive glasses, together with soluble silica, leads to rapid, heterogeneous nucleation of bone mineral within the osteoid and thereby new bone forms rapidly on Class A bioactive materials. [34] A Class A bioactive implant is osteoproductive as well as osteoconductive due to both intracellular and extracellular interactions with newly forming bone. [26, 34] A Class B bioactive implant, such as HA, is only osteoconductive due to extracellular interactions with osteoid.

These findings have profound implications with respect to the future of biomaterials. They indicate that the genetic basis for cell proliferation may be controlled by an alloplastic material. This is especially important, if true, since it is generally agreed that the decrease in trabecular bone volume, and associated degradation in strength, which occurs with aging and osteoporosis is due to a continual decrease in rate of osteoblast proliferation. [35] The delicate balance between osteoblast-based bone formation and osteoclast-controlled bone resorption is shifted towards resorption with increasing age and the bone trabeculae become progressively thinner and weaker (figure 17). The basic science studies summarized in figure 16 indicate that the degradative process of osteopenia and osteoporosis may be reversed, or perhaps even prevented, by therapeutic treatments or dietary intake of biologically active silicon. Increased research on the biochemical basis for Class A bioactive behavior is needed. When the mechanisms are established for the enhanced bone repair illustrated in figures 8 and 9 it may be possible to design bioactive compositions specific for repair of osteoporotic bone and enhance the quality of bone bonded to a bioactive implant.

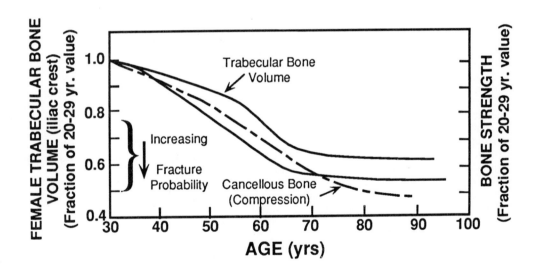

Figure 17 Effect of age on trabecular bone volume and strength.

ACKNOWLEDGMENT

The author gratefully acknowledges financial support of Air Force Office of Scientific Research (Grant No. F49620-92-J-0351).

References

1) Hench, L.L., Splinter, R.J., Allen, W.C., and Greenlee, T.K., Jr.: *J.Biomed.Mater.Res.,* 1972, **2,** 117.
2) Hench, L.L.: *J.Amer.Ceram.Soc.,* 1993, **74,** 1487.
3) *An Introduction to Bioceramics*, eds. Hench, L.L. and Wilson, J.: 1993, World Scientific Publishing, NY.
4) *Handbook of Bioactive Ceramics, Vol. I: Bioactive Glasses and Glass-Ceramics*, eds. Yamamuro, T., Hench, L.L., and Wilson, J.: 1990, CRC Press, Boca Raton, FL.
5) *Handbook of Bioactive Ceramics, Vol. II: Calcium Phosphate and Hydroxylapatite Ceramics*, eds. Yamamuro, T., Hench, L.L., and Wilson, J.: 1990, CRC Press, Boca Raton, FL.
6) Gross, U., Kinne, R., Schmitz, H.J., and Strunz, V.: *CRC Critical Reviews in Biocompatibility,* 1988, **4,** 2.
7) Ham, A.W.: *Histology*, 1969, J.B. Lippincott, Philadelphia.
8) Hench, L.L. and Paschall, H.A.: *J.Biomed.Mater.Res.,* 1974, **5,** 49.
9) Clark, A.E., Pantano, C.G., and Hench, L.L.: *J.Amer.Ceram.Soc.,* 1976, **59,** 37.
10) Ogino, M., Ohuchi, F., and Hench, L.L.: *J.Biomed.Mater.Res.,* 1980, **14,** 55.
11) Seitz, T.L., Noonan, K.D., Hench, L.L. and Noonan, N.E.: *J.Biomed.Mater.Res.,* 1982, **16,** 195.
12) Wilson, J., Pigott, G.H., Schoen, F.J., and Hench, L.L.: *J.Biomed.Mater.Res.,* 1981, **15,** 805.
13) Hench, L.L., Paschall, H.A., Allen, W.C., and Piotrowski, G.: *National Bureau of Standards Special Publication,* 1975, **415,** 19.
14) Hench, L.L. and Ethridge, E.C.: *Biomaterials: An Interfacial Approach*, 1982, Academic Press, NY.
15) Jarcho, M.: *Clin.Ortho.Relat.Res.,* 1981, **157,** 259.
16) de Groot, K.: *Bioceramics of Calcium-Phosphate*, 1983, CRC Press, Boca Raton, FL.
17) de Groot, K., Klein, C.P.A.T., Wolke, J.G.C., and de Blieck-Hogervorst, J.: p. 3 in *Handbook of Bioactive Ceramics, Vol. II: Calcium Phosphate and Hydroxylapatite Ceramics*, eds. Yamamuro, T., Hench, L.L., and Wilson, J.: 1990, CRC Press, Boca Raton, FL.
18) Williams, D.F.: p. 43 in *Biocompatibility of Tissue Analogs*, Vol. II, ed. Williams, D.F.: 1985, CRC Press, Boca Raton, FL.
19) Cook, S.D.: in *Bioceramics*, Vol. 3, eds. Hulbert, J.E. and Hulbert, S.F.: 1992, Rose-Hulman Institute of Technology, Terre Haute, IN.
20) Jarcho, M., Kay, J.I., Gummaer, R.H., and Drobeck, H.P.: *J. Bioengr.,* 1977, **1,** 79.
21) Kokubo, T.: *J.Non-Cryst.Sol.,* 1990, **120,** 138.
22) Kitsugi, T., Yamamuro, T., and Kokubo, T.: *J. Bone and Joint Surg.,* 1989, **71A,** 264.
23) Yoshii, S., Kakutani, Y., Yamamuro, T., Nakamura, T., Kitsugi, T., Oka, M., Kokubo, T., and Takagi, M.: *J.Biomed.Mater.Res.,* 1988, **22,** 327.
24) Wilson, J., Yu, L.T., and Beale, B.S.: p. 139 in *Bioceramics 5*, eds. Yamamuro, T., Kokubo, T., and Nakamura, T.: 1992, Kobonshi Kankokai, Kyoto, Japan.
25) Oonishi, H., Kushitani, S., Yasukawa, E., Kawakami, H., Nakata, A., Koh, S., Hench, L.L., Wilson, J., Tsuji, E., and Sugihara, T.: p. 139 in *Bioceramics 7*, eds. Andersson, O.H. and Yli-Urop, A.: 1994, Butterworth-Heinemann, Oxford, England.
26) Wilson, J. and Low, S.B.: *J. Appl. Biomaterials,* 1992, **3,** 123.
27) Kawanabe, K., Yamamuro, T., Nakamura, T., Kokubo, T., Yoshihara, S., and Shibuya, T.: p. 233 in *Bioceramics 5*, eds. Yamamuro, T., Kokubo, T., and Nakamura, T.: 1992, Kobonshi Kankokai, Kyoto, Japan.
28) Ducheyne, P. and McGucken, J.F., Jr.: p. 175 in *Handbook of Bioactive Ceramics, Vol. II: Calcium Phosphate and Hydroxylapatite Ceramics*, eds. Yamamuro, T., Hench, L.L., and Wilson, J.: 1990, CRC Press, Boca Raton, FL.
29) Doyle, C.: p. 195 in *Handbook of Bioactive Ceramics, Vol. II: Calcium Phosphate and Hydroxylapatite Ceramics*, eds. Yamamuro, T., Hench, L.L., and Wilson, J.: 1990, CRC Press, Boca Raton, FL.
30) Soltez, U.: p. 137 in *Bioceramics: Materials Characteristics Vs In Vivo Behavior*, Vol. 523, eds. Ducheyne, P. and Lemons, J.E.: 1988, Annals NY Academy of Sciences, NY.
31) Bonfield, W.: p. 173 in *Bioceramics: Materials Characteristics Vs In Vivo Behavior*, Vol. 523, eds. Ducheyne, P. and Lemons, J.E.: 1988, Annals NY Academy of Sciences, NY.

32) Keeting, P.E., et al.: *J. Bone and Mineral Res.*, 1992, **7**, 1281.
33) Vrouwenvelder, W.C.A., Groot, C.G., and de Groot, K.: *J.Biomed.Mater.Res.*, 1993, **27**, 465.
34) Hench, L.L.: p. 3 in *Bioceramics 7*, eds. Andersson, O.H. and Yli-Urop, A.: 1994, Butterworth-Heinemann, Oxford, England.
35) Revell, P.A.: p. 183 in *Pathology of Bone*, 1986, Springer-Verlag, Berlin.

Materials Science Forum Vol. 293 (1999) pp. 65-82
© 1999 Trans Tech Publications, Switzerland

Novel Biomedical Materials Based on Glasses

T. Kokubo

Division of Material Chemistry, Faculty of Engineering, Kyoto University,
Sakyo-ku, Kyoto 606-01, Japan

Keywords: Bioactive Glasses, Bioactive Glass-Ceramics, Bone-Bonding Mechanism, Bioactive Metals, Bioactive Polymers, Bioactive Cements, Hyperthermia Treatment, Radiotherapy

ABSTRACT: Some bioactive glasses and glass-ceramics have occupied important positions in bone-repairing materials. Fundamental studies on their bone-bonding mechanism indicate that Si-OH and Ti-OH group on their surfaces induce the nucleation of bonelike apatite which is resposible for bonding to living bone. These findings provide us techniques for forming the biologically active bonelike apatite even on metals and organic polymers. A bioactive composite cement also can be derived from bioactive glass-ceramics. Ferrimagnetic crystals-containing glass-ceramics are useful as thermoseeds for hyperthermia treatment of cancers. Chemically durable glasses containing Y or P at high level can play an important role in radiotherapy of cancers.

INTRODUCTION

Since discovery of Bioglass by Hench et al in early 1970's, various kind of glasses and glass ceramics have been found to bond to living bone. Some of them have already occupied important positions in bone-repairing materials. Bone-bonding mechanism of them has been also fundamentally revealed considerably. These findings enable us to design even bioactive metals, polymers and composite cements. On the other hand, it has been also shown during the last decade that glass-based materials can play an important role in cancer treatments such as hyperthermia treatment and radiotherapy. Recent progress in these area is reviewed in the present paper.

BIOACTIVE GLASSES AND GLASS-CERAMICS IN CLINICAL USES

Hench et al discovered in early 1970's that some glasses in the system Na_2O-CaO-SiO_2-P_2O_5 spontaneously bond to living bone [1]. Since then, various kind of bioactive glasses [2] and glass-ceramics [3-6] have been developed. Some of them have already occupied important positions in bone-repairing materials. For example, Bioglass® in the system Na_2O-CaO-SiO_2-P_2O_5 and related glasses are already clinically used as artificial middle ear bones, periodontal implants, and maxillofacial implants [7], because of their high bioactivities. Glass-ceramic A-W containing crystalline apatite $(Ca_{10}(PO_4)_6(O, F_2)$ and wollastonite $(CaO \cdot SiO_2)$ are now clinically used as artificial vertebrae, intervertebral spacers, iliac spacers, bone fillers etc. [8] by the name of Cerabone® A-W, because of its high mechanical strength (215 MPa in bending strength) as well as high bioactivity. More than 5,000 patients have received it as their bone substitutes during the last 3 years, since approval for sales from Ministry of Welfare, Japan in 1991. Figure 1 shows some examples of artificial bones of Cerabone® A-W. When this artificial bone substitutes for a vertebra of a sheep, it is soon bonded to the surrounding cancerous bone, as shown in Fig. 2.

Even Cerabone® A-W, however, can not replace highly loaded bones such as femoral and tibial bones, since its fracture toughness (2 MPa $m^{1/2}$) is not so high as that (6 MPa $m^{1/2}$) of human cortical bone and its elastic modulus (118 GPa) is not so low as that (30 GPa at maximum) of human cortical bone. For these purposes, metallic implants coated with bioactive glasses [9] or hydroxyapatite [10] by physical methods are being used. Thus formed bioactive layers are not stable in the living body for a long period. Tough material itself is desired to bond to a living bone. In order to obtain such material, bone bonding mechanism must be revealed.

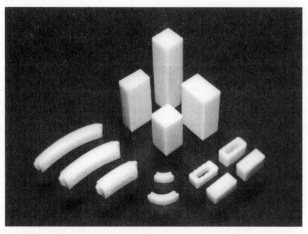

Fig. 1. Iliac spacers (left), artificial vertebrae (middle top), spinous process spacers (middle bottom) and intervertebral spacers (right) of Cerabone®A-W.

Fig.2. Contact microradiograph of Cerabone® A-W which substituted for a vertebra of a sheep and
was bonded to the surrounding cancerous bone.

BONE-BONDING MECHANISM

All the bioactive glasses and glass-ceramics hitherto known bond to living bone through
an apatite layer which is formed on their surfaces in the living body, as shown in Fig. 3 [11,12].
This apatite layer can be reproduced on the surfaces of the bioactive glasses and glass-ceramics
even in an acellular simulated body fluid (SBF) with ion concentrations (Na^+ 142.0, K^+ 5.0, Mg^{2+}
1.5, Ca^{2+} 2.5, Cl^- 147.8, HCO_3^- 4.2, HPO_4^{2-} 1.0 and SO_4^{2-} 0.5 mM) nearly equal to those of human
blood plasma. According to thin film X-ray diffraction and Fourier transform infrared reflection
spectroscopy [13, 14], the surface layer consists of a carbonate-containing hydroxyapatite ($Ca_{10-x}[(CO_3$
$, HPO_4)_x(PO_4)_{6-x}](OH)_{2-x}$) with Ca/P atomic ratio less than 1.67. These structural and compositional
characteristics are very similar to those of the apatite in the bone. Consequently, bone-producing
cell, osteoblast, can preferrentially proliferate on this apatite layer and differentiate to produce the
apatite and collagen, as shown in Fig. 4 [15]. These biological apatite and collagen soon fully
occupy the space on the bioactive material and come to direct contact with the surface apatite
layer, as shown in Fig 3. When this occurs, a tight chemical bond is formed between the bone
apatite and the surface apatite, as proved by the observation that a couple of Bioglass and Cerabone®
A-W are spontaneously bonded together in SBF through an apatite layer formed at their interface
[16]. It is, therefore, believed that the essential condition for an artificial material to bond to living
bone is the formation of the apatite layer on their surfaces in the body. The problem to be revealed
is what kind of material can form the bonelike apatite layer on their surfaces in living body.

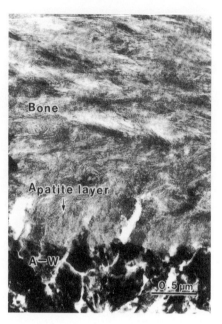

Fig. 3. Transmission electron micrograph of interface of Cerabone®A-W with rat tibia.

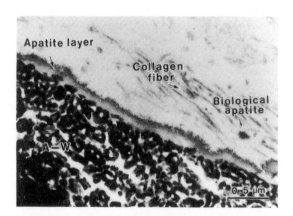

Fig. 4.   Transmission electron micrograph of collagen fibers and apatite crystals produced by osteoblast on the surface of apatite layer on Cerabone®A-W.

MECHANISM OF APATITE FORMATION

It has been believed for a long period that the P_2O_5 is the essential component for the bioactive material, since the bioactive materials started with Bioglass containing P_2O_5. The present authors, however, later showed that even P_2O_5-free Na_2O-CaO-SiO_2 ternary glasses form the bonelike apatite layer on their surfaces in SBF, as shown in Fig. 5 [17]. It should be noted that even Na_2O-SiO_2 and CaO-SiO_2 binary glasses can form the bonelike apatite layer. It is already confirmed in vivo that the CaO-SiO_2 binary glasses can bond to living bone through the bonelike apatite layer [18]. In the CaO-SiO_2-P_2O_5 ternary system, formation of the bonelike apatite layer is limited to CaO, SiO_2-based compositions, but not to the CaO, P_2O_5-based compositions, contrary to the conventional expectation, as shown in Fig. 6 [19]. This is explained as follows [20]. The CaO, SiO_2-based glasses release the calcium ion mainly into SBF, whereas the CaO, P_2O_5-based glasses release the phosphase ion mainly. Since both these ions are components of the apatite, the release of these ions increases the ionic activity product of the apatite in the surrounding fluid almost equally, as shown in Fig. 7. Despite it, the ionic activity is soon suddenly decreases only in the case of CaO, SiO_2-based glasses. This is because the apatite is formed on their surfaces by consuming the calcium and phosphate ions from the fluid. This indicates that the surfaces of the CaO, SiO_2-based glasses provide favorable sites for the apatite nucleation. The CaO, SiO_2-based glasses form a hydrated silica layer prior to the formation of the apatite in SBF. This means that the hydrated silica induces the apatite nucleation. This is proved by the observation that when a pure silica gel prepared by a sol-gel method is soaked in SBF, it forms the bonelike apatite on it, as shown in Fig. 8 [21]. The same silica gel, however, does not form the apatite, after it is heated above 900 ℃ [22]. This means that a kind of silanol group on the surfaces of CaO, SiO_2-based glasses is responsible for the apatite nucleation. The calcium ion released from them accelerates the apatite nucleation. Once the apatite nuclei are formed, they spontaneously grow by consuming the calcium and phosphate ions from the surrounding fluid, since the body fluid is already supersaturated with respect to the apatite.

Then, our interest is extended to the question "Is there any kind of other material which can induce the apatite nucleation?" Recently, we found that a pure titania gel prepared by a sol-gel method also induces the apatite nucleation, as shown in Fig. 9 [23]. This indicates that alkali titanate-based glasses also form the bonelike apatite on their surfaces in living environment. Actually, it was found that some alkali titanate-based glasses in the system K_2O-TiO_2-SiO_2 form the bonelike apatite on their surfaces in SBF, as shown in Fig. 10 [24]. These findings provide us novel techniques for forming the bonelike apatite even on metals with high fracture toughness and organic polymers with low elastic modulus.

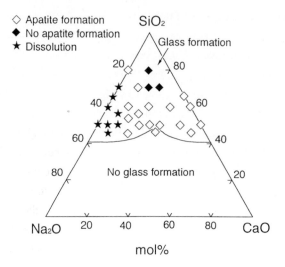

Fig. 5. Apatite formation on glasses in the system $Na_2O-CaO-SiO_2$ in SBF within 30 days.

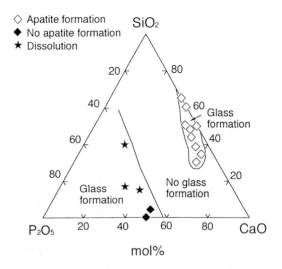

Fig. 6. Apatite formation on glasses in the system $CaO-SiO_2-P_2O_5$ in SBF within 30 days.

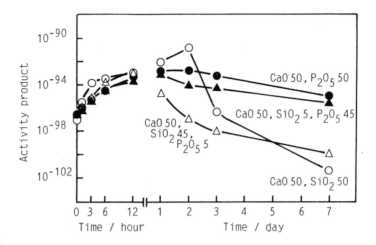

Fig. 7. Variation of ionic activity product of the apatite in SBF due to soaking of glasses.

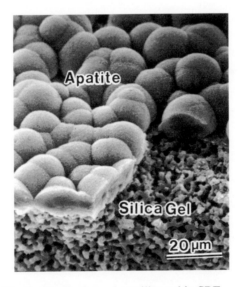

Fig. 8. Apatite formed on silica gel in SBF.

Fig. 9. Apatite formed on titania gel in SBF.

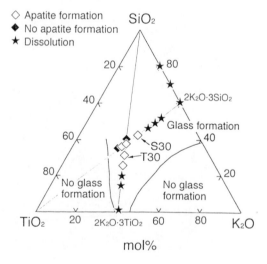

mol%

Fig. 10. Apatite formation on glasses in the system K₂O-SiO₂-TiO₂ in SBF within 30 days.

APATITE FORMATION ON METALS

It was shown in the previous section that alkali titanate-based phase can form the bonelike apatite on their surfaces in SBF. The alkali titanate layer can be formed on titanium metal and its alloys such as Ti-6Al-4V and Ti-6Al-2Nb-Ta by an alkali treatment. For example, when they are exposed to 10M-NaOH solution at 60 ℃ for 24 h, their passive layer of titanium oxide reacts with it to form an alkali titanate hydrogel layer, as shown in Fig. 11 [25]. This gel layer is stabilized as an amorphous alkali titanate layer by a heat treatment around 600 ℃. When thus treated metals are implanted into living body or soaked in SBF, the bonelike apatite layer is spontaneously formed on their surfaces, as shown in Fig. 12. In this reaction, the alkali ion in the amorphous layer exchanges with hydronium ion in the surrounding fluid. As a result, hydrated titania is formed to induce the apatite nucleation, and the released alkali ion accelerates the apatite nucleation by increasing the ionic activity product of the apatite in the surrounding fluid due to increase in pH, as shown in Fig. 11. Thus formed apatite layer is strongly bonded to the substrates, since it is integrated with the substrates through the hydrated titania and titanium oxide layers gradually changed.

It can be seen from this example that a tough material which can exhibit bioactivity by itself can be obtained by utilizing an amorphous phase as a nucleating agent for apatite. Thus obtained bioactive metals are believed to be useful as artificial bones even under high load-bearing conditions such as hip joint, since they have high fracture toughness as well as high bioactivity. Elastic moduli of these metallic implants are, however, appreciably higher than that of human cortical. As a result, surrounding bones are liable to be resorbed, because of stress shielding by the metals. In order to obtain bioactive materials with lower elastic modulei, the bonelike apatite must be grown on organic polymers.

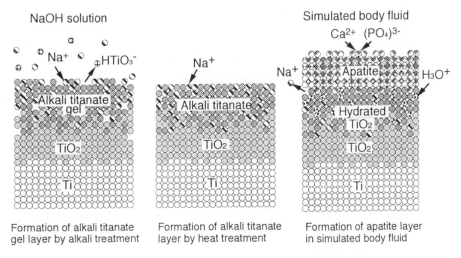

Fig. 11. Schematic representation of reactions at surface of Ti metal on alkali treatment, heat treatment and soaking in SBF.

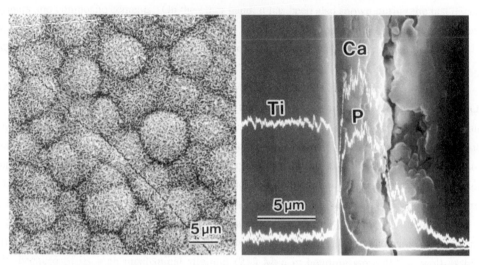

Fig. 12. Scanning electron micrograph (left) of a surface of Ti metal which was soaked   in SBF
after the alkali and heat treatments, and electron prove X-ray microanalysis (right) of its
cross section.

APATITE FORMATION ON POLYMERS

Not only solid silica gel but also a silicate ion which is dissolved from the silica gel and adsorbed on the surface of an organic polymer induces the apatite nucleation in SBF [26]. These findings provide us a following biomimetic process [27] for forming the bonelike apatite layer on polymer substrates. First, a polymer substrate is placed on granular particles 150 to 300 μ m in size of a CaO-SiO$_2$-based glass, e.g., a glass of the composition MgO 4.6, CaO 44.7, SiO$_2$ 34.0, P$_2$O$_5$ 16.2, CaF$_2$ 0.5 wt%, soaked in SBF, as shown in Fig. 13. Then the polymer substrate is soaked in another solution highly supersaturated with respect to the apatite, e.g. a solution (1.5 SBF) with ion concentrations 1.5 times those of SBF. During the first treatment, a silicate ion which was dissolved from the glass particles and adsorbed on the surface of the polymer substrate induces the apatite nucleation there. The calcium ion released from the glass particles accelerates the apatite nucleation. During the second treatment, the apatite nuclei spontaneously grow by consuming the calcium and phosphate ions from the surrounding fluid in situ. When sufficiently large number of the apatite nuclei are formed during the first treatment, a dense and continuous bonelike apatite layer is formed on any kind of organic polymers during the second treatment, as shown in Fig. 14. The period of the first treatment required for forming the continuous apatite layer during the second treatment, which is defined as an induction period for the apatite nucleation, is almost 1 day for most of the polymers including poly(ethylene terephthalate)(PET), poly(methyl mechacrylate)(PMMA), polyeher sulfone(PESF), polyamide 6 (Nylon 6), poly ethylene (PE) and poly(tetrafluoro ethylene)(PTTE) [27].

The thickness of the apatite layer increases linearly with increasing time of the second treatment at a rate of 1.7 μ m/ day at 36.5 ℃ as shown in Fig. 15 [28]. The growth rate of the apatite layer increases with increasing temperature of the second treatment up to 7 μ m/ day at 60 ℃. The growth rate also increases with increase in ion concentrations of the solution, as well as with shaking of the solution for the second treatment. These results indicate that the growth of the apatite layer is controlled by mass transport across the interface between the apatite and the solution [28].

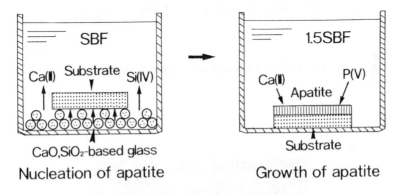

Fig. 13. A biomimetic method for forming apatite layer on organic polymer substrate.

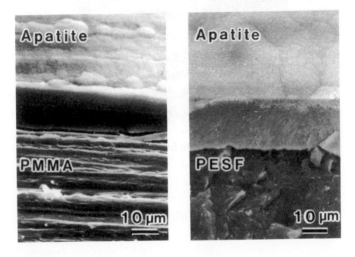

Fig.14. Dense and uniform apatite layer formed on organic polymers.

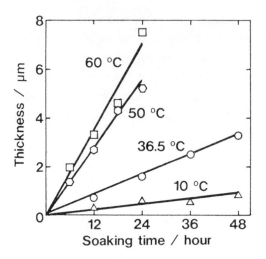

Fig. 15. Thickness of apatite layer as a function of period of the second treatment at various temperatures.

The adhesive strength of the apatite layer to the substrate largely varies with the kind of polymers, i.e. 4.40, 3.48, 1.93, 1.06, 0.63 and $<1.1 \times 10^{-2}$ MPa for PESF, PET, PE, PMMA, Nylon 6 and PTFE. These values are remarkably increased by grow-discharge pretreatment of the polymers in oxygen gas, up to 9.55, 9.77, 7.51, 5.76, 7.03 and 2.14 MPa for PESF, PET, PE, PMMA, Nylon 6 and PTTE, respectively [29]. These increases are attributed to the formation of polar groups such as carbonyl, ester, hydroxyl and carboxyl ones on the surfaces of the polymers by the glow-discharge treatment. They might effectively trap the silicate ion in the first treatment and also form a bond with calcium and/or hydroxyl ions of the apatite in the second treatment.

The bonelike apatite layer can be formed uniformly not only on flat surfaces but also on curved surfaces of individual fine fibers constituting a fabric, as shown in Fig. 16 [28]. The structure of this apatite-polymer fiber composite is very similar to that of the natural bone, where apatite is deposited on collagen fibers, at least partly. Therefore, it is expected that if this kind of composite could be fabricated into a three dimensional structure analogous to that of the natural bone, the resultant material could exhibit analogous mechanical properties to those of the natural bone, not only in its fracture toughness but also elastic modulus, as well as high bioactivity. This type of composite is believed to have great potential as bone-repairing material.

Before coating   After coating

Fig. 16 PET fine fibers constituting a fabric before coating (left) and after (right).

BIOACTIVE-COMPOSITE CEMENT

For fixing nonbioactive metallic and polymeric implants to the surrounding bones or for filling bone defects in complex shapes, a cement which shows enough fluidity to be formed into any shapes or injected within a few minutes and then solidifies to show a high mechanical strength is needed. Hitherto, PMMA cement has been mainly used for this purpose. It, however, does not bond to living bone, and hence its fixation is liable to loosen. Various attempts for obtaining bioactive cements have been made by mixing crystalline calcium phosphates [30, 31] or glassy calcium silicate [32] powders with aqueous solutions. The solidified cements are, however, poor in their mechanical strengths.

It was recently found that a mixture of glass-ceramic A-W powders with some resin gives bioactive cement with high mechanical strength [33]. Glass-ceramic A-W powders, which were pretreated with a silane coupling agent, is first mixed with 0.4 wt% of dibenzoyl peroxide powders as a polymerization initiator. On the other hand, bisphenol-a-glycidylmethacrylate (Bis-GMA) liquid is mixed with triethylene glycol dimethacrylate (TEGDMA) liquid by 1:1 in weight ratio, and added with 0.2 vol% of N,N-dimethyl-p-toluidine as a polymerization accelerator. Then, the former powder is mixed with the latter liquid 70:30 in weight ratio in order to form a cement. The mixed paste shows enough fluidity to be shaped into any shapes or injected within a few minutes, and then solidifies. When it is implanted into a bone defect before solidification, some monomers are released from the surface of the cement to expose A-W powders to the surrounding body fluid, as shown in Fig. 17. Consequently, A-W powders react with the body fluid to form the bonelike apatite on the surface of the cement as well as intergranular spaces of the A-W particles, which were formed by partial release of the monomers. Soon the cement bonds to living bone through the apatite layer. The solidified cement gives a bending and compressive strengths as high as 135 and 270 MPa, respectively, both of which are about twice those of PMMA cement. Its fracture toughness and Young's modulus are 1.6 MPam$^{1/2}$ and 14 GPa, respectively.

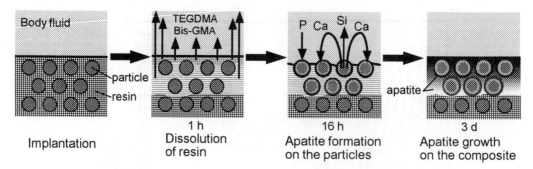

| | 1 h | 16 h | 3 d |
| Implantation | Dissolution of resin | Apatite formation on the particles | Apatite growth on the composite |

Fig. 17. Schematic representation of reactions at surface of A-W—Bis-GMA/TEGDMA cement in living body.

GLASS-CERAMICS FOR HYPERTHERMIA TREATMENT OF CANCER

Glass-ceramics containing ferrimagnetic or ferromagnetic crystalline phases in a nearly innert or bioactive matrix are useful as thermoseeds for hyperthermia treatment of cancers. They are compatible to the living tissue. When they are implanted around malignant tumors and placed under an alternating magnetic field, the tumors are locally heated up to temperatures above 43 ℃ effective for cancer treatment by their magnetic hysteresis losses. A glass-ceramic containing lithium ferrite ($LiFe_5O_8$) in a Al_2O_3-SiO_2-P_2O_5 glassy matrix precipitating small amount of hematite (α-Fe_2O_3) [34], and that containing magnetite (Fe_3O_4) in a CaO-SiO_2 glassy matrix precipitating wollastonite [35] have been developed for this purpose. It is already shown by animal experiments that they are effective for cancer treatment as shown in Fig. 18 [36]. Further increase in efficiency of heat generation is, however, still investigated.

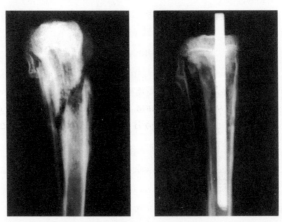

Fig. 18 X-ray photographs of rabbit tibia 5 weeks after transplantation with bone tumore. Left: no treatment, Right: a ferrimagnetic glass-ceramic pin was inserted and placed under an alternating magnetic field for 50 min.

GLASSES FOR RADIOTHERAPY OF CANCER

Chemically durable glasses containing Y or P at high level are useful as radioactive seeds for in situ radiation of cancers. They can be activated to emitter of β-ray with short half-life time by neutron bombardment. When they are injected to malignant tumors through the vascular system, in a form of microspheres 20 to 30 μm in size, after activated, they are entrapped in the capillary bed of the tumors and give large local radiation doses of the short ranged highly ionizing β-ray with little irradiation to the neighboring normal tissues. The radioactive element is hardly released when the chemical durability of the glasses is high. Their radioactivities are rapidly decayed to the negligible level after the cancer treatment, because of their short half-life times.

A glass of the composition Y_2O_3 40, Al_2O_3 20, SiO_2 40 wt% shows high chemical durability and can be easily formed into microspheres by the conventional melt-quenching technique. Therefore, this is already clinically used for radiotherapy of liver and kidney cancers [37]. The half-life time of radioactive element ^{90}Y in this glass is, however, as short as 64.1 h and hence its radioactivity may result in substantial decay even before the cancer treatment. The ^{31}P with 100% natural abundance is also activated to β-emitter, ^{32}P, with a little longer half-life time of 14.3 d by neutron bombardment. The biological effectiveness of ^{32}P is about 4 times that of ^{90}Y. It is, however, difficult to prepare chemically durable glasses containing P at high level by the conventional melt-quenching technique. The present authors recently showed that such glass can be prepared by ion implantation technique. When the P^+ ion is implanted into a chemically durable silica (SiO_2) glass at energy as high as 200 keV, it is entrapped as phosphorus colloids in the glass giving the maximum concentration at about 200 nm in depth, by doses as large as $1 \times 10^{18} cm^{-2}$. As a result, a chemical durable glass containing large amount of P can be obtained [38], even if the surface structure of the glass is damaged by the ion implantation. Thus obtained P^+-implanted silica glass is believed to be useful as a radioactive seed for in situ radiation of cancers.

CONCLUSION

Glass has large variety in its composition and structure. It can be also crystallized. It is, therefore, believed to have great potential in biomedical fields as in other fields. New kind of novel biomedical materials are expected to be derived from glasses in future.

REFERENCES

1) Hench, L.L., Splinter, R.J., Allen, W.C. et al: J. Biomed. Mater. Res. Symposium, 1971, 2 (1) 117.

2) Andersson, Ö. H., Karlsson, K.H., Kangasniemi, K. et al: Glastech. Ber. 1988, 61, 300.

3) Brömer, H., Pfeil, E. and Kos, H.H.: German Patent No. 2, 326,100, 1973.

4) Kokubo, T., Shigematsu, M., Nagashima, Y. et al: Bull. Inst. Chem. Res., Kyoto Univ., 1982, 60, 260.

5) Höland, W., Naumann, J., Vogel, W. et al: Wiss, Z. Friedrich Schiller Univ. Jena Math. Naturwiss, Reihe, 1983, 32, 571.

6)   Berger, G., Sauer, F., Steinborn, G. et al: in Proceedings of XIV Intern. Congress. Glass, Vol, 3a, Mazarin, O.V. (Ed), Nauka, Leningrad, 1989, p.120.

7)   Wilson, J., Yli-Urpo, A. and Risto-Pekka: in Introduction to Bioceramics, Hench, et al (Ed), World Scientific, Singapore, 1993, p.63.

8)   Yamamuro, T.,: in Introduction to Bioceramics, Hench, L.L. et al (Ed), World Scientific, Singapore, 1993, p.89.

9)   Hench, L.L. and Andersson, Ö.: in Introduction to Bioceramics, Hench, L.L. et al (Ed), World Scientific, Singapore, 1993, p.239.

10) Lacefield, W.R.: in Introduction to Bioceramics, Hench, L.L. et al (Ed), World Scientific, Singapore, 1993, p.223.

11) Kokubo, T.: in Bone-bonding biomaterials, Ducheyne et al (Ed), Reed Healthcare Communications, Netherland, 1993, p.31.

12) Neo, M., Kotani, S., Nakamura, T. et al: J. Biomed. Mater. Res., 1992, 26, 1419.

13) Kokubo, T., Ito, S., Huang, Z.T. et al: J. Biomed. Mater. Res., 1990, 24, 331.

14) Kokubo, T., Kushitani, H., Sakka, S. et al: J. Biomed. Mater. Res., 1990, 24, 1721.

15) Neo, M., Nakamura, T., Yamamuro, T. et al: J. Biomed. Mater. Res., 1993, 27, 999.

16) Kokubo, T., Hayashi, T., Sakka, S. et al: Yogyo-Kyokai-Shi, 1987, 95, 785.

17) Kim, H-M., Miyaji F., Kokubo T., et al: J. Am. Ceram. Soc., 1995, 78, 2405.

18) Ohura, K., Nakamura, T., Yamamuro, T. et al: J. Biomed. Mater. Res., 1991, 25, 357.

19) Ohtsuki, C., Kokubo, T., Takatsuka, K. et al: J. Ceram. Soc. Japan, 1991, 99, 1.

20) Otsuki, C., Kokubo, T. and Yamamuro, T.: J. Non-Crystl. Solids, 1992, 143, 84.

21) Li, P., Ohtsuki, C., Kokubo, T. et al: J. Am. Ceram. Soc., 1992, 75, 2094.

22) Cho, S.B., Nakanishi, K., Kokubo, T. et al: J.Am. Ceram. Soc., 1995, 78, 1769.

23) Li, P., Ohtsuki, C., Kokubo, T. et al: J. Biomed. Mater. Res., 1994, 28, 7.

24) Kim, H-M., Miyaji, F., Kokubo, T., et al, unpublished data.

25) Miyaji, F., Zhang, X., Yao, T. et al: in Bioceramics Vol. 7, Anderson, Ö.H. et al (Ed), Butterworth-Heinemann, Oxford, 1994, p.119.

26) Cho, S.B., Miyaji, F., Kokubo, T. et al: in Bioceramics Vol. 8, Wilson, J. et al(Ed), Pergamon, Oxford, 1995, p.493.

27) Tanahashi, M., Yao, T., Kokubo, T. et al: J. Am Ceram. Soc., 1994, 77, 2805.

28) Hata, K., Kokubo, T., Nakamura, T., et al: J. Am. Ceram. Soc., 1995, 78, 1049.

29) Tanahashi, M., Yao, T., Kokubo, T. et al: J. Biomed. Mater. Res., 1995, 29, 349.

30) Chow, L.: J. Ceram. Soc. Japan, 1991, 99, 954.

31) Constanz, B. R., Ison, I.C., Fulmer, M.T. et al: Science, 1995, 267, 1796.

32) Kokubo, T., Yoshihara, S., Nishimura, N. et al: J. Am. Ceram. Soc., 1991, 74, 1739.

33) Kawanabe, K., Tamura, J., Yamamuro, T. et al: J. Appl. Biomater., 1993, 4, 135.

34) Luderer, A.A., Borrelli, N.F., Panzarino, J.N. et al: Radiation Res., 1983, 94, 190.

35) Ebisawa, Y., Sugimoto, Y., Hayashi, T. et al: J. Ceram. Soc. Japan, 1991, 99, 7.

36) Ikenaga, M., Ohura, K., Yamamuro, T. et al: J. Orthopaedic Res., 1993, 11, 849.

37) Day, D.E. and Day, T.E.: in Introduction to Bioceramics, Hench, L.L. et al (Ed), World
    Scientific, Singapore, 1993, P. 305.
38) Kawashita, M., Miyaji, F., Kokubo, T. et al: in Bioceramics Vol. 8, Wilson, J. et al (Ed),
    Pergamon, Oxford, 1995, P. 501.

Materials Science Forum Vol. 293 (1999) pp. 83-98
© 1999 Trans Tech Publications, Switzerland

Biomimetic Assembly of Nanostructured Materials

M. Sarikaya, H. Fong, D.W. Frech and R. Humbert

Materials Science and Engineering, University of Washington, Seattle, WA 98195, USA

Keywords: Biomimetics, Genetics, Hierarchy, Hybrid Materials, Self-Assembly

1. ABSTRACT

Biomimetics, simply, is materials science and engineering through biology. It involves studying microstructures and mechanisms of organismal tissue formation, correlation of processes and structures with physical and chemical properties, and using this knowledge-base to design and synthesize technological materials for practical engineering and health applications. Therefore, it firstly involves the ways biological organisms synthesize and form complex materials (soft and hard tissues) via self- and co-assembly of organic macromolecules that are ion-carriers, templates, and growth modifiers controlling microfabrication of highly ordered, hierarchical inorganic structures with multifunctional properties. Secondly, it allows one to develop novel genetic and materials engineering strategies of processing new and advanced materials with tailored structures and properties for practical technological applications.

2. WHAT IS BIOMIMETICS?

Materials are a major key for industrialization, as witnessed throughout the human endeavour from stone age, to iron age, to steel and silicon ages. During the last couple of decades, aided by the advent of nano- and molecular scale imaging and testing techniques, it became clear that new materials with controlled discontinuities from the nanometer to the macro scale display unprecedented physical (optical, magnetic, semiconducting, and mechanical) properties.[1-4] The ability to produce such complex (multiphasic and nanostructured) materials depends largely on the developing ways to form microstructures with controlled phase compositions and morphologies, and in significant quantities.[5-7] However, as the structural variations in materials get smaller, their preparation becomes more complex and require sophisticated instrumentation. Although many physical properties are improved and the materials produced are used in specialized applications, many of these techniques are energy inefficient. With all the successes in producing these materials structures and their applications, there are still significant issues involved in the uniformity of size, crystal structure, morphology, chemistry of the phases formed that are difficult to control and require sophisticated instrumentation; in addition they are not easily amenable to scale-up to allow wide-scale applications, and, in many instances, toxic byproducts are produced.

Living organisms, on the other hand, produce biological materials with properties that surpass those of the currently produced engineering materials.[8,9] The main reason for this is that the structures of these biomaterials are highly organized at all scales of dimension from molecular to the nanometer, to the micrometer, and to the macro scale, and often in a hierarchical manner.[5,10-12] These materials are, simultaneously, "smart," dynamic, complex, self-healing, and multifunctional having, e.g., optical, mechanical, electrical, and magnetic properties. It is known that these intricate structures are built-up in aqueous environments at ambient conditions, under the close scrutiny of organic macromolecules that not only collect and transport raw constituents, but also self- and co-assemble them into short- and long-range ordered substrates and nuclei, and direct them to grow into intricate nanoarchitectures that make up myriad different tissues of biological organisms from bacteria to the human body.[10-14]

The biological materials systems of technological interest, such as hard tissues (including bone, teeth, small particles, and mollusk shell structures) and soft tissues (including spiders' silks, **mucus**, and biomembranes) can be a rich source of inspiration for design and synthesis concepts for developing novel synthetic materials.[6,10,13] The goal of this new field, i.e., *biomimetics*, at the cross-roads of physical (materials science, physics, chemistry) and biological sciences (microbiology, biochemistry, and genetics), is to eventually synthesize novel technological materials based on biological structures. The recently developed concepts and their practical outcome as new class(es) of materials called as smart, intelligent, functionally-gradient, hierrachical, nanotechnological or nanostructured, hybrid, complex, biotechnological, or molecular materials depending on one's discipline; [1-4] but they are all included in the paradigm of biomimetic materials. Therefore, from physical scientists' point-of-view, biomimetics is materials science and engineering through biology. This paper discusses the fundamental ideas behind biomimetics, an emerging interdisciplinary field involving the facscinating worlds of nanoscience at the cross-section of biological and physical sciences.

3. RESEARCH IN BIOMIMETICS

Molecularly engineered inorganic materials with structural order at the nanoscale (such as thin films and small particles) are of significant practical importance in applications from optical coatings to microelectronics. Production of synthetic materials with controlled structures at molecular and nanoscale that exhibit physical properties with engineering interest has grown significantly in recent years.[12,15-20] These include membrane-mimetic inorganic particles (formation within vesicles and micelles and on organic mono- and bi-layers) [15] surfactant-aided hydrothermal particles, self-assembled multilayer organic-inorganic structures within electrolytes, [16] inorganic materials formed within or on ordered polymeric composites, and water-surfactant-preceramic molecular systems that result in mesoporous structures.[17] The common denominators in these systems are the involvement of an organic "template," room temperature synthesis in mostly aqueous environments, nanoscale control of the microstructures, and long-range order, similar to the conditions in biological systems during biomineralization. References to biomineralization and bioself-assembly is, therefore, common in these synthetic systems.[4-20] It is not yet fully clear, however, what specifically the set of fundamental phenomena are that govern the detailed event(s) that successively or concurrently take place during transport, molecular self-assembly, synthesis, growth, and systems formation processes that result in the extraordinary structures and physical/chemical properties that we encounter in the biological systems. In the following section, several such biological hard-tissue are described as examples to bring about the issues involved.

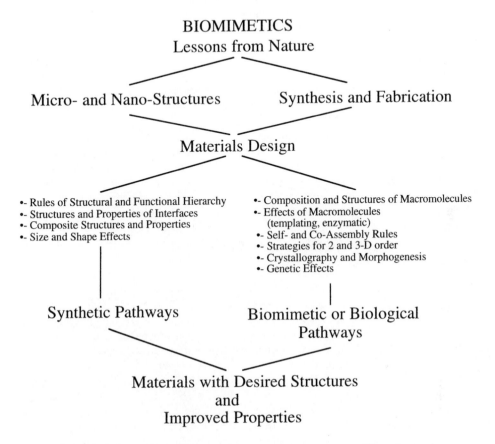

Figure 1 - The concept of biomimetics and the ways to assemble and organize ordered structures to achieve functional materials

3.1 EXAMPLES OF BIOLOGICAL HARD TISSUES

Synthesizing via the rules of biomineralization may be a way of producing novel materials with better structural control and, hence, better properties than those of the existing technological materials because ogranisms function under genetic control and synthesize consistently reproducable materials.[4-13] In biomineralized structures formation of a usually crystalline mineral phase is modified or controlled by proteins and possibly by other macromolecules. Both the chemical composition (stoichiometry) and crystallography of the inorganic components are characteristic of and controlled by the organism.[20-24] For example, several isomorphic forms of the crystals may be present within different tissues of a given organism, demonstrating the complexity of the organismal control over biomineralization. In red abalone *(Haliotis rufescens)*, the prismatic and nacreous sections contain calcitic (rhombohedral) and aragonitic (orthorhombic) isomorphs of $CaCO_3$, respectively (see details next subsection) and microstructures of these sections are unique to this species among all mollusks, and they are not encountered in geological or synthetic $CaCO_3$.[22] On the other hand, nautilus (*Nautilus pompelius*) is a cephalopod and it has a different overall shape of the shell since its habitat is significantly different than that of abalone. Even so, it has a shell that consists of outer prismatic and inner nacreous layers similar that of abalone. In fact, the nacre section, consisting of layered platelets of aragonite separated by organic matrix, is difficult to distinguish from that of abalone (compare figures 2(a) and 3(B)).

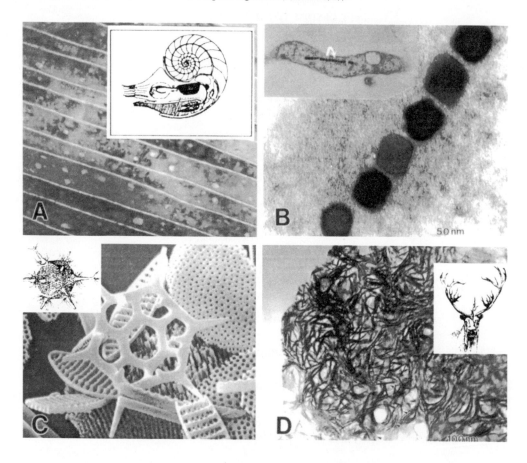

Figure 2 - (a) An example of layered biocomposite: TEM image of the nacreous section of nautilus (inset) showing aragonite platelets separated by a thin-film of organic matrix. (b) An example of particulate bioceramic: magnetite (Fe_3O_4) formation in magnetotactic bacteria: Aquaspirillum magnetotacticum. Magnetite particles are arranged as a string; they are single crystalline, single domained, and highly perfect (inset) (TEM image). (c) Example of Porous-amorphous structures: skeletal units of radiolaria: SEM image displays various radiolaria skeletal with highly symmetric shapes in a diatomaceous earth sample. (d) Example of three-dimensional matrix with ultrafine inorganic precipitates: antler bone from caribou (inset). TEM image shows elongated particles are hydroxyapatite in a collageneous matrix.

Another example is ultrafine magnetite (Fe_3O_4) particles that are formed within magnetotactic bacteria, such as *Aquaspirillum magnetotacticum*. This example also illustrates the specificity and complexity of biological structures that vastly exceed that of synthetic systems (figure 2(b)).

The shape of the magnetite particles in this bacterium species is species-specific, but their size in all species is within the single magnetic domain range (400-600Å diameter) and, hence, the particles are perfect single crystals. To achieve similar materials specifics, more stringent conditions are required for the industrial production of magnetite (or ferrite and chromite which are also used for magnetic storage media).[26]

Similar complex, highly intricate, and hierarchical structures are also encountered in hard tissues such as sea-urchin skeletal units, bone and dentin, and insect cuticles.[6] In the case of radiolaria (as well as in many disordered hard tissues, such as sponge spicules), the mineral phase is amorphous silica (i.e., no long-range order of the atomic species, but only short range ordered, glassy, structures); yet the organization and overall shape of the biomaterial is highly symmetric and it contains hierarchically-porous architecture (figure 2(c)).[8,9] In this case, the organic matrix (if any, containing proteins) is probably incorporated into the mineral phase forming a complex molecular composite material, not too different from the newly discovered synthetic molecular sieves (or mesoscopic porous structures) containing water-surfactant-silica components.[27-29]

Finally, the structures of bone and antler-bone are similar consisting of a three-dimensional network of protein/polysaccharide network that act as the scaffolding for the formation of ultrafine (30 Å thick and 500 Å long) crystalline particles of hydroxyapatite (calcium phosphate). Both of these tissues, which are both hard and tough, contain up to 65% mineral phase. Figure 2 (d) shows a TEM image of antler bone (from caribou) displaying an intricate nanostructure. Similar to mammalian tooth dentin [30], the bones have truly hierarchical structures starting with various types of collagen leading to tropocollagen associated with HA particulates at the nanometer dimension through many stages of hierarchy leading to a complex structure of bone that is tailored depending on the position in the body and its function.[31]

3.2 A MODEL LAYERED CERAMIC-BIOPOLYMER COMPOSITE: NACRE OF MOLLUSK SHELLS

A biological hard tissue that has highly ordered, relatively simpler (compared to bone and dentin) structures that result in multifunctional mechanical properties is shells of mollusk species, and, in particular, nacre.[10,22a-b] Many mollusks, such as gastropods (e.g., *Haliotis rufescens*, red abalone), cephalopods (e.g., *Nautilus pompelius*, chambered nautilus) and bivalves (e.g., *Pinctada margaritafera*, pearl oyster) have two-tiered structures across their shells; exterior prismatic and interior nacre structures (Figure 3-H).[8,9] The prismatic section has calcite crystallites (rhombohedral $CaCO_3$ - calcium carbonate - crystal structure with P3m space group) that are ordered in columnar form, perpendicular to the plane of the shell, having a base length of few micrometers and length of 5-10 μm. The nacre structure (also called "mother-of-pearl") has aragonite crystallites (orthorhombic $CaCO_3$ unit cell with Pmmm space group) that have pseudohexagonal platelike shape that are ordered in "brick and mortar" structure figure 3 A-D).[22a-b] The platelets have a thickness of about 0.25 μm (up to 0.8 μm in pinctada), and edge length about 2-10 μm. Each section, prismatic and nacre, has about 2-4 vol.% organic matter (consisting, of proteins, and probably, polysaccharides and lipids) that, as a thin (5-10 nm-thick) film, surrounds the inorganic particles. The organic matrix self-assembles and acts as a multifunctional template in controlling overall formation and the hierarchical architecture of the inorganic component including nucleation and mineralogy of the individual $CaCO_3$ particles, their crystallography with respect to each other and the organic matrix, their growth into defined shapes, internal (lattice) and external (particle) morphology, and overall shape of the shell.[20-24]. When eventually formed, the shell as a system becomes a "perfect armor," a protective device for the organism,[25]as it has been for over 500 million years. [8,9]

The extraordinary mechanical properties, a combination of fracture toughness and strength (figure 3-J), compared to those of high technology ceramics, mainly orginate from the shell architecture, its texture as an anisotropic composite material (aragonite inorganic layers and thin organic matrix), and interface properties between these components. The high tortuosity seen on the fracture surface of nacre (figure 3-G) can only account for, perhaps, a fractional increase in fracture toughness but not all (actual increase in fracture toughness is 20-40 times that of polycrystalline $CaCO_3$ that is synthetically prepared). Detailed fractographic analysis (by SEM) in the vicinity of microindentations on nacre samples in the edge-on orientation (i.e., indentor parallel to the flat surface, i.e., (001), of the platelets) indicates that two energy absorbing mechanisms are in operation. Firstly, if the resolved stresses are shear (i.e., parallel to the platelet plane, then the platelets slide on each other (figure 3 -E), where sliding is provided by the nanolayerd nature of the organic film (which is composite of 5 layers).[32] Secondly, if the applied stresses are normal, then the crack opening (via the separation of the successive platelets) is prevented by the bridging of the platelets by the organic matrix which superplastically deforms to form ligaments between the platelets. The combination of fracture toughness (measured in 3-point bending) and fracture (bend) strength (four-point bend) of nacre structures range upward from 4 through 11.5 Mpa-m$^{1/2}$, and up to 15 MPa-m$^{1/2}$ for abalone[33] and pinctada,[34] respectively. These properties of the nacre structure, which is a model biological ceramic-polymer (cerpoly) layered composite, are much higher than those of high performance structural ceramics, such as Al_2O_3, ZrO_2, Si_3N_4, SiC, and B_4C, and at the level of the most successful ceramic-metal composites (cermets), such as Al_2O_3-Al, B_4C-Al and WC-Co containing more than 80% vol. ceramic component (not 95%!).[35]

Although a full understanding of the correlations have not yet been developed, these desirable engineering properties could result from a combination of structural and compositional factors, including nano- and micro(layered)-architecture, crystallography, size and shape of the component phases, their nanoscale and microscale (size-dependent) properties, and coupling of the inorganic and organic phases across the interfaces. It is, therefore, tempted to predict what the properties would be if advanced ceramics are used (instead of $CaCO_3$) in the design of novel cerpoly or cermet composites with layered architecture. The result of one such attempt is shown in figure 3-K (L is the schematic figure), where layered B_4C-Al composite was achived by thin films of Al. Compared to 3-dimensional network of B_4C-Al, the laminated cermet shows marked increase (about 30-40%) (figure 3-J). However, this increase is an order of magnitude smaller than what was achived in the model biomaterial. Part of the reason for this inadequacy in traditional processing approach, probably, are due to, firstly, the limitations in current processing technique (tape casting) which allows a layer thickness not less than 10 µm, which is far larger than 0.25 µm in nacre; secondly, due to lack of understanding the effects of nanoarchitectures of the component phases; and thirdly, limited of understanding of coupling across the interface.

Besides offering nacre as an excellent model for a layered ceramic based material, the mollusks also provide the whole-shell as a model system for a material-device, namely as an (impact resistant and energy absorbing) armor. There are many examples of biological hard tissues that can be biomimetic models for design of engineering materials. For example, mammalian teeth are dynamic wear-resistant cutting and grinding tools;[36] magnetite particles in bacteria are components of a natural compass;[37] sea-urching skelatal units (such as spines) are a rare example of precipitation-hardened ceramics ($MgCO_3$ within $CaCO_3$),[38] sponge spicules can behave as excellent natural fiber-optic materials.[39] In the case of mollusk shells, the two-tiered structure could be thought as a system, provides impact resistance against the predators, then the outer prismatic section should be hard (to stop the outside projectile, such as a rock or a beak of a predator) and inside should be compliant enough to absorb the energy produced during the impact. In fact, as shown in Table-I, prismatic section has a higher hardness* than that of nacre (215 versus 158 Kg/mm^2). However, in the geological counterparts, single crystal aragonite has a much higher hardness than that of calcite, in the same direction. This means that the organism has reversed the mechanical properties of the sectional composites to provide a hard surface and a tough inlay in producing this evolutionary very successful biosystem, an ideal biological armor.[25]

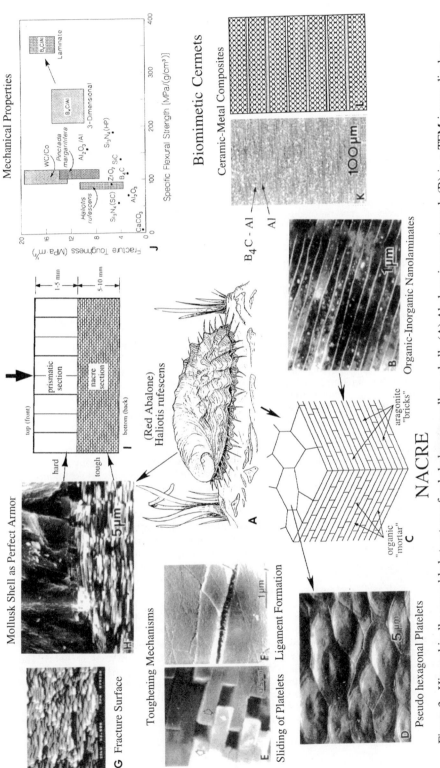

Figure 3 - Hierarchically assembled structure of red abalone, a mollusc, shell. (A) Abalone is a gostropod; (B) is a TEM image displays layered structure of aragonite (CaCO₃)/organic matrix in edge-on view; (C) reveals the brick-and-mortar architecture of nacre; (D) is an SEM image of cleaved nacre displaying pseudo-hexagonal aragonite platelets in face-on view; (E) and (F) are SEM images showing the sliding and ligament formation, respectively, in the vicinity of a microindentation; (G) displays tortuous fracture surface of nacre; (H) fracture surface across nacre/prismatic interface and (I) is its schematic illustration. (J) is mechanical properties chart of biocomposites, advanced structural ceramics and cermets. (K) is an light optical image of layered cermet and (L) is its schematic illustration.

Table - I: Microhardness of CaCO₃ Minerals and Biocomposites

Material (001) plane	Vickers Microhardness (Kg/mm^2)	Knoop Microhardness (Kg/mm^2)
Geological Aragonite	447	340
Mineral Calcite	174	177
Biogenic Aragonite (Nacre)	158	135
Biogenic Calcite (Prismatic)	215	210

The challenge in biomimetics of nacre structure in mollusk shells, therefore, would be to mimick its hierarchical structure at the nano-. micro, and macro-architectural levels using a hybrid material involving inorganic/organic phases, such as oxide, carbonate, or carbide hard ceramics and synthetic or macromolecular soft polymeric (or a metal) phase, mimicking all aspects of the hierarchy and interfaces.[4,22]

3.3   ENAMEL BIOMIMETICS

Another example for functional biological tissue with self-assembled, but complex microstructure, is mammalian dental tissues. For example in human tooth, there are two hard tissues: enamel on the outer section (constituting almost all ceramic with hydroxylapatite, HA, being the inorganic material) and dentin on the inner section (about equal amounts of organic, collegenous matrix, and HA crystallites) with an interfacial region called dentin-enamel junction (DEJ) [40,41].   The DEJ is the region where structural interlocking takes place and which provides mechanical coupling between the hard (enamel) and tough (dentin) regions forming a functionally-gradient "biogenic tool" [42]. Since enamel is a non-living tissue, its regenaration hardly ever been tried, and its (partial) replacement has traditionally been performed by "external" fillers (metallic, ceramic, polymeric, or their cocomposites), each with various successes but never replacing the permanent tissue. It is desirable, therefore, that regeneration of enamel would be a permanent solution.

In spite of the fact that structure of the enamel has been relatively well categorized, essential details are still lacking. For example, the details of the structure of dental-enamel junction (DEJ) is very limited, yet, as discussed above, DEJ serves as the "bridge" connecting two functionally diverse portion of the tooth material. The spatial distribution of the organic and inorganic phases in dentin and enamel at this transition zone appears to be necessary in elucidating formation of enamel and the effects of dentin in its development. Tooth is a "tool" having certain functional characteristics.[36] Local structures of the enamel, their correlation throughout the volume of the hard tissue, as well as its shape and internal morphology are necessary to understand its function as a tool. The knowledge how dentin and enamel are associated with each other structurally is essential in terms of understanding the biomechanical properties of the tooth as a functional composite materials system.

Preliminary experiments performed in this group have focused on the structural and mechanical coupling at DEJ.[43] Figure 4 shows images from a DEJ region of a juvenile human tooth using atomic force microscopy (AFM) and transmission electron microscopy (TEM). The AFM image (figure 4(b)) reveals both enamel (gray) and dentin regions (light). The first among several important structural features is that the interphase region is not a simple and narrow plane but one that displays a tortuous, interpenetrating dentin and enamel phases (Figure 4(b)). There are subtle differences in morphology, size, and crystallography of hydroxyapatite

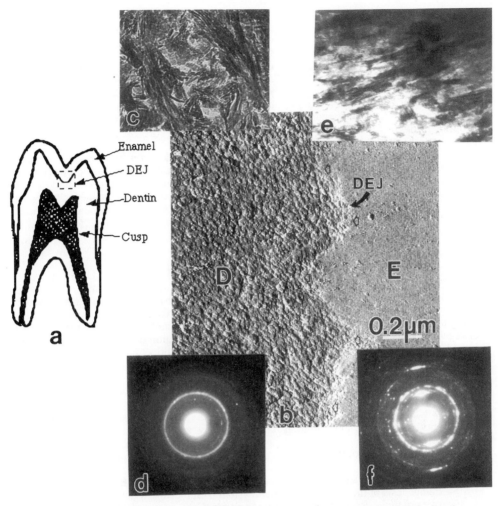

Figure 4 - (a) Schematic cross-section of a tooth displaying positions of the hard tissues. (b) AFM image of the dentin-enamel junction; (c) TEM image of the dentin, and (d) its electron diffraction pattern (diffraction rings are an indication of randon orientations of HA crystallites). (e) TEM image of enamel and (f) its diffraction pattern (diffraction arcs indicate textured orientations of HA crystallites).

(HA) crystallites in dentin and enamel sides of the interface. HA crystallites in dentin are elongated platelets with 3 nm thickness and 30-50 nm edge length (c), cystallographically randomly oriented (d - inset) within the collageneous matrix. In enamel, the HA crystallites are much larger and longer (e), columnar shaped with 25-30 nm diam. and 20/1 aspect ratio, and textured (f - inset). The arcs in this SAD pattern indicates that the HA crystallites are highly oriented perpendicular to the interface plane. These structural features are an indication that while properties are highly isotropic in dentin, these are unisotropic in enamel.

Because of the size of the associated tissues and interface region, local mechanical properties were determined using a nanomechanical testing instrument[*] (using a Berkovich tip at

250 µNewton maximum load). Force-depth (F-d) curves were obtained from dentin and enamel; a single crystal HA mineral was used as a control sample. Comparable to those reported in the literature [44,45] the hardness values range from 6.5 ± 1.1 GPa for fully dense mineralogical apatite, 3.53 ± 0.51 GPa for enamel to 0.45 ± 0.07 GPa for dentin. The large difference in hardness could be attributed to large size of the crystals and their high packing density in enamel compared to those in dentin. Because of the texture in enamel, it is expected that further variations are present in anisotropic mechanical properties. Further measurements are underway to obtain hardness across and along the interface, in dentin and enamel including local crystallographic and morphological analyses to fully understand structure-property coupling through DEJ in human teeth.[43].

A detailed knowledge of architecture of both enamel and dentin, and their local properties in the vicinity of DEJ, including (anisotropic) three-dimensional structure and property maps at nano- and micrometer scale, are essential for potential replacement and regeneration of enamel both as a material and a mechanical system. This could be done biomimetically by using biocompatible (synthetic) polymeric scaffolds within which HA crystallites could be formed in a controlled manner using proteins (such as enamalogenin and taftelin) as nucleators and habit formers.[42]

4. MAJOR ISSUES OF STUDY IN BIOMINERALIZED TISSUES

Despite numerous studies in the field (mostly by non-physical scientists) the fundamental understanding of biomineralization has not yet been clearly explained and the understanding of inorganic phase formation at organic interfaces has been nominal. Some of the fundamental issues in biomineralization and inorganic material formation at organic interfaces in organisms include:[21,22]

- Mechanisms of inorganic and organic ion transport,
- Assembly processes and pattern formation of the organic macromolecules,
- Temporal and spatial nucleation of the inorganic phase(s),
- Growth and long-range order of the inorganic phase(s) and interfaces with organics,
- Origins of specificity in mineralogy, crystallography, and morphology of inorganic particles,
- Structural hierarchy of the (organic/inorganic) hybrid material (both in 2- and 3-dimensional space),
- Structural coupling between hard tissues, and between hard and soft tissues
- Morphogenesis of the biogenic materials system (i.e., biological device or tissue),
- Genetics of biological materials formation and function.

This list can easily be extended to include many (mostly unknown) finer steps, including complex physiological processes, that the organism is involved in the build-up fo the materials system. Two such important factors are, first, proteins (or macromolecules, in general) that are used as templates, or catalyzers, for the inorganic phase formation, and (possibly myriad of) enzymes that are involved in the transport of ions, their assembly with the macromolecules, switching of the mineralogy, directed amorhpous or crystalline growth.

Furthermore, fundamental aspects, such as the differences in intra- and extra-cellular biomineralization strategies have to be fully addressed. The importance of the components of the organic matrices and templates, such as proteins, polysaccharides, and lipids, have to be clearly identified. These include their functionalities, locations, detailed structures, interactions, self- and co-assembly processes with other organic macromolecules, such as enzymes and inorganic species and phases.

Similarly, in the synthetic systems, the issues related to fundamental interactions that bring the inorganic ion and the micro- or macromolecule together, and basic differences in co-, self- and forced-assembled processes that form the overall, nanoscale ordered systems, have only recently started to be addressed. Therefore, we are far away from reaching the fundamental insights governing the self-assembly process in biological organisms and utilize these as design-guidelines in synthetic materials systems. The example below gives one such recent successful attempt from this group in controlled assembly of inorganic particles on engineered polypeptides, a technique which is likely to lead many applications in nano- and molecular-materials technology.

5. GENETIC ENGINEERING APPROACH TO MATERIALS PROCESSING: ASSEMBLY OF NANOPARTICLES USING ENGINEERED POLYPEPTIDES

In biological microstructures described above, self-organized molecular architectures (proteins) provide functionalized surfaces with affinity toward the inorganic component(s). These surfaces may reduce the energy for nucleation, induce a specific mineralogy, orient the crystals, or control the growth via inhibition. Nanostructured material could be developed for use in electronic, photonic, and chemical applications using macromolecules through self- and co-assembly procedures [12,16-19] because of the consistency and precision that can be achieved through biomimetics pathways.[7] These macromolecules may be extracted from biological composites, they may be *de-novo* designed, or synthetic counterparts may be used. Alternatively, as discussed here, engineered polypeptides which recognize inorganic surfaces (ceramics, metals, and semiconductors), as well as numerous biological surfaces, can be isolated from combinatorial genetic libraries.[46]

Combinatorial genetic techniques permit isolation of specific recognition elements for surfaces, including those not realized by natural proteins, in the absence of *apriori* prediction of necessary structures.[47] Here we demonstrate the controlled assembly of nanometer-scale gold particles on functionalized spherical (and flat) surfaces that are used as substrates in aqueous solutions using engineered gold-binding proteins [46] as recognition elements.[48] Selection of gold binders was designed to produce modular, independently folding metal recognition sequences. Gold-binding sequences were isolated as extracellular loops of maltoporin which were subsequently fused to the amino terminus of alkaline phosphatase with retention of gold-binding activity.[46] Many proteins bind tightly to gold at low salt concentrations; however our engineered proteins were selected to bind at higher salt concentrations. As a result, they exhibit substantially improved binding compared to the native *E. coli* alkaline phosphatase even in the presence of detergent. We used binders with 5-, 7-, and 9- tandem repeats of a 14-amino acid motif in our experiments (figure 5).[48] The 7-repeat gold binder (GB) was used for most assembly experiments. It should be emphasized that the binding motif does not contain cysteine which is known to form a covalent thiol linkage with gold, the linkage to the gold surface in self-assembled monolayers.[18]

Figures 5(b and c) summarize the protocol in which submicrometer-diameter polystyrene (PS) beads were first functionalized by glutaraldehyde ($HCO(CH_2)_3CHO$). Subsequently, beads were derivatized with GB or control alchalinephosphatase (AP), incubated with preformed colloidal gold (using the Turkevich technique)[49] and washed thoroughly.[47] Glutaraldehyde activated PS beads or PS beads derivatized with control (e.g., bovine serum albumin) proteins resulted in low to absent gold binding activity (Figure 5(c)). When beads are coated with GB-protein, preformed gold-particles decorated the surfaces of the beads (Figure 5(d)). This image is what is basically a demonstration of quantum dot structures assembled on a biomimetically engineered polypeptides.[50]

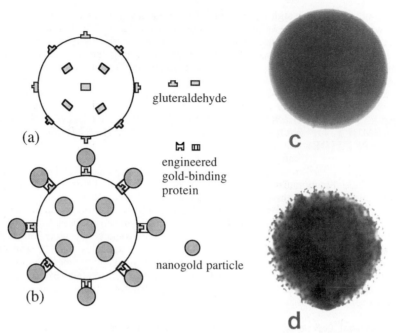

Figure 5 - (a) and (b) Diagram of protein-dependent attachment of colloidal gold to polystyrene beads, with and without the protein, respectively. (c) TEM image of the bead activated with glutaraldehyde with control-protein (bovine serum albumin) present before incubation with nanogold; the image shows almost nonexitent gold particle attachment to the non-specific protein. (d) Nanogold particles attached to polystyrene surface via 9-repeat; gold particles are assembled on surface-specific engineered proteins via a templating mechanism (gold particles are about 12 nm diameter).

Our experiments are the first in demonstrating that inorganic materials can be assembled at the nanoscale by proteins that have been genetically-engineered to bind to selected materials surfaces. These results could lead to new avenues in nanotechnology[1-3,18,19] in tailoring the formation and assembly of ordered structures of metals, functional ceramics, semiconductors, and ferroelectrics through formation, shape-modification, and assembly of materials, and development of surface-specific protein coatings (in biotechnology). The combinatorial genetic approach is a general one which should be applicable to numerous surfaces.[50]   The modularity of binding motifs should allow genetic fusion of peptide segments recognizing two different materials that would allow co-assembly into nanocomposite structures.

6.   FUTURE DIRECTIONS IN BIOMIMETICS

Many questions in biomimetics listed in Section 3 need to be addressed in systematic ways by a collaborative research among the physical and biological scientists because of the requirement of diverse expertise in this new field. The understanding of some of the most fundamental issues are critical to the success of biomimetics in micro- and nano-fabricating novel high technology, multifunctional, and smart materials at the molecular levels under less stringent, more environmentally friendly, and energy efficient pathways including genetic engineering strategies.

For instance, although significant work is being done in the topical area, an understanding of interface structure with the organics and assembly mechanism(s) of inorganic phases are fundamental in deciphering the templating or enzymatic mechanism(s) that are central to many biological hard tissue formation. On the one hand, it is necessary that surface-specific macromolecules are engineered via genetic and biochemistry strategies; these polysaccharides and DNA, as well as proteins, would have inorganic-surface activity, can assemble into many different geometrical configurations, and would act as templates directing materials nucleation, mineral switch, growth, and/or assembly. On the other hand, it is also essential to develop synthesis and processing strategies to form inorganic (such as ceramic) materials (thin films, small particles, porous structures) in aqueous solutions with pH between 4 - 10 and at ambient temperatures (<100 °C), conditions similar to those in biological systems. Biomimetics would then provide the means to marry these two parallel approaches.

The effort to accomplish the complex task of successfully designing and processing materials with bioinspired ways should involve many researchers with a wide ranging expertise from materials sciences, crystal formation, microscopic and spectroscopic characterization, to solution and polymer chemistry, surface physics, microbiology, orthopedics, biochemistry, bioengineering, biotechnology and to genetics (figure 6). Once this cross-collaboration is established among multidisciplinary fields, the highly exciting field of biomimetics would yield significant results that will benefit all disciplines of science, and will produce new technological materials for the benefit of all humankind, most likely during the better part of the next century.

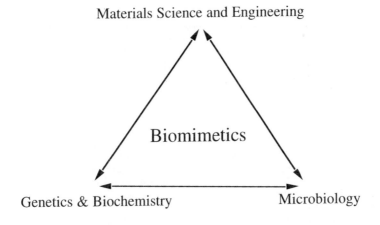

Figure 6 - A schematic illustration of marriage of materials (physical) sciences and biological sciences in the creation of biomimetics field.

7. ACKNOWLEDGMENTS:

This work was supported by US-AFOSR and -ARO. Our collaborators (Professors Stanley Brown, Clement Furlong, and James Staley, and Dr. Jun Liu) are gratefully acknowledged.

8. REFERENCES

[1] Siegel, R.W., 1993, *Physics Today*, **46**, 64.
[2] National Research Council, *Materials Science and Engineering for the 1990's: Maintaining Competitiveness in the Age of Materials* (NAS, Washington, D.C., 1989).
[3] Drexler, K. E., 1991, *Nanotechnology*, **2**, 113.
[4] *Hierarchically Structured Materials*, Proc. of MRS, Vol. 220, Aksay, I. A., Baer, E., Tirrell, D., and Sarikaya M. (eds.) (Materials Research Society, Pittsburgh, 1992). Also, other MRS Proceedings, such as Vols. 174 and 330.
[5] *Hierarchical Structures in Biology as a Guide for New Materials Technology*, National Materials Advisory Board, National Research Council, Vol. NMAB-464 (National Academy Press, Washington., DC, 1994).
[6] *Biomolecular Materials*, Proc. of MRS, Voil. 330, Viney, C., Case, S.T., and Waite, J. H. (eds.) (Materials Research Society, Pittsburgh, 1993).
[7] *Biomimetics: Design and Processing of Materials*, Sarikaya, M. and Aksay, I. A. (eds.) (American Institute of Physics, New York, 1995).
[8] *Biomineralization*, K. Simkiss and K. M. Wilbur (Academic Press, New York, 1989).
[9] *On Biomineralization*, Lowenstam, H., and Weiner, S. (Oxford U. Press, New York, 1989).
[10] Currey, J. D., 1987, *J. Mater. Edu.* **9** [1-2] 118.
[11] Calvert, P. and Mann, S., 1988, *J. Mater. Sci.*, **23**, 3801.
[12] Ozin, G. A., 1992, *Adv. Mat.*, **4**, 612.
[13] *Biomimetic Materials Chemistry*, Mann, S. (ed.) (VCH, New York, 1996).
[14] Degens, T., 1976, *Top. Curr. Chem.*, **64**, 1.
[14] Addadi, L. and Weiener, S., 1985, *Proc. Nat. Acad. Sci.*, USA, **82**, 4110.
[15] *Membrane-Mimetic Approach to Advanced Materials* , Fendler, J. H., Advances in Polymer Science Series, Vol. 113 (Springer and Verlag, Berlin, 1994).
[16] Mingming, F., Chy Hyung, K, Sutorik, A. C., Kaschak. D. M., and Mallouk. T. E., 1996, in: *Amorphous and Crystalline Insulating Thin Films*, Vol. 199, Warren, W, L., Devine, R. A. B., Matsumura, M., Cristoloveanu, S., Homma, Y., and Kanicki, J. (eds.) (Materials Research Society, Pittsburgh) pp. 377-81.
[17] See, for example, Stucky, G. et al.: Monnier, A. et al., 1993, *Science*, **261**, 1299.
[18] Whitesides G. M., Mathias J. P., and Seto C. T., 1991, *Science*, **254**, 1312.
[19] Knoll W. et al., 1993, *Syn. Metals.*, **6**, 5.
[20] Weissbuch, I., Addadi, L., Lahav, M., & Leiserowitz, L., 1991, *Science,* **253**, 637.
[21] Sarikaya, M. and Aksay, I. A., in: Structure, *Cellular Synthesis, and Assembly of Biopolymers*, S. T. Case (ed.) (Springer-Verlag, Berlin, 1992) pp. 1-25.
[22] Sarikaya, M., Liu, J., and Aksay, I. A., in: *Biomimetics: Design and Processing of Materials*, Sarikaya, M. and Aksay, I. A. (eds.) (American Institute of Physics, New York, 1995) pp. 34-85.
[23] Wierzbicki A., Sikes C. S., Madura J. D., and Drake B. 1994, *Calcif Tissue Int*, **54**, 133.
[24] Belcher A. M. et al., 1996, *Nature*, **381**, 56.
[25] Shapiro, B. and Sarikaya, M., 1998, to be published in *J. Mater. Res.*
[26] Magnetic Recording Materials, *MRS Proce'edings*, 1990, Vol. XV, No. 3, March.
[27] Kreske, C. T. *et al.*, 1992, *Nature*, **359**, 710.
[28] Stucky, G. D. *et al.*, 1994, *Mol. Cryst. and Liq. Cryst.*, **240**, 187.
[28] Liu., J. *et al.*, 1996, *Science*, **69**, 131.
[30] Lees, S. & Rollins, F. R., 1996*J. Biomechanics*, **5**, 557.
[31] Glimcher, J., in: *The Chemistry and Biology of Mineralized Tissues: Wulf's Law Revisted*, Veis, A. (ed.) (Elsevier, New York and Amsterdam, 1981). pp. 617-673.
[32] Weiner, S. and Traub, W., *Phil. Trans. R. Soc., Lond. B.*, 1984, **304**, 425.

[33] Gunnison, K. E., Sarikaya, M., Liu, J., andAksay, I. A., in *Hierarchically Structured Materials Symp. Proc.* (Materials Research Society, Pittsburgh, PA 1992) pp. 171-184.

[34] Sawyer, S. and Sarikaya, M., unpublished (1994).

[35] Davidge, W, *Mechanical Behavior of Ceramics* (Cambridge University Press, Cambridge, 1979).

[36] Fox, P. G. 1985, *J. Mater. Sci.* **15**, 3113.

[37] *Iron Biominerals*, Blakemore, R. B. and Frankel, R. (eds.) (Pergamon, New York, 1992).

[38] Liu, J. and Sarikaya, M., unpublished (1992).

[39] CattaneoVietti-R. *et al.*, 1996, *Nature*, **383**, 397.

[40] Simmer, J. P. and Funcham, A. G., *Crit. Rev. Oral. Biol. Med.*, **6** (2) 84-108 (1995).

[41] Robinson, C., Brookes, Steven J., Shore, Roger C., and Kirkham, J., 1998, *Eur. J. Oral Sci.*, **106**, 282.

[42] Snead, M. L., University of Southern California, Private Communication.

[43] Fong, H., Research in progress (1998).

[44] Kinney, J. H. *et al.*, 1996, *J. Biomech. Eng.*, **11**,133.

[45] Willems, G. *et al.*, *J. Biomeh. Mater. Res.*, **27** (1993) 747.

[46] Barbas III, C. F., Rosenblum, J. S. and Lerner, R. A., Proc. 1993, *Natl. Acad. Sci., USA*, **90**, 6383.

[47] Brown, S.1997, *Nature Biotechnol.* **15**, 269.

[48] Humbert, R., Brown, S., and Sarikaya, M., 1998, submitted.

[49] Turkevich J., Stevenson P. C., and Hillier J., 1951, *Trans. Faraday. Soc. Disc.*, **11,** 55.

[50] Brown, S. and Sarikaya, M., work in progress (1998).

Materials Science Forum Vol. 293 (1999) pp. 99-106
© *1999 Trans Tech Publications, Switzerland*

Bioceramics - Current Status and Future Trends

J.F. Shackelford

Department of Chemical Engineering and Materials Science, University of California,
Davis, CA 95616, USA

Keywords: Bioceramics, Biomaterials, Biomimetics, Ceramics, Glasses, Hydroxyapatite

ABSTRACT

A wide range of ceramic materials are currently used in biomedical applications, from chemically pure oxides, such as alumina, to the chemically and microstructurally complex ceramic-matrix composites. Among the most significant applications are the hydroxyapatite coatings on hip prostheses and the use of hydroxyapatite and/or related calcium salts for bone defect and fracture repair. Increasing applications of ceramic and glass materials can be expected in both orthopaedics and dentistry. The use of these materials in cancer treatment also appears promising. The development of biomimetic processing techniques and protein delivery systems represent the cutting edge in bioceramics, involving the merger of materials science with the biological sciences.

THE RANGE OF CERAMIC MATERIALS CURRENTLY USED IN BIOMEDICINE

In this volume, we have seen a range of ceramics with a corresponding range of biomedical applications. In the opening chapter dealing with the historical development of bioceramics, we found that these materials can be conveniently classified by their chemical reactivity. [1] Alternately, bioceramics can be classified by their primary chemical constituents, from those composed of relatively simple oxides to others which are relatively complex chemically. [2] Such a classification system is given in Table 1. For the glass-ceramics and ceramic-matrix composites, microstructural features are often a significant factor also.

As noted in the opening chapter, *simple oxides* were central to the development of modern bioceramics in the late 1960's. [3] Dense, high-purity (> 99.5 %) sintered Al_2O_3 structural ceramics were the first bioceramics widely used in a clinical setting. [4] They are used for orthopaedic surgery as hip prostheses and in dentistry as dental implants, based on a combination of good strength, modest fracture toughness, high wear resistance, good biocompatibility and excellent corrosion resistance. Experience with alumina in orthopaedic surgery for more than twenty years demonstrates its high degree of biocompatibility, including minimal scar formation which would prevent the mechanical bonding of bone to a porous implant surface. Although well-engineered alumina/alumina ball-and-socket hip prosthesis systems in Europe have demonstrated successful long-term performance, the lack of the highest quality control standards can lead to severe problems with wear debris damage. In the United States, hip prosthesis designs are largely confined to the use of alumina for the ball, with the socket being made from ultra high molecular weight polyethylene (UHMWPE). A variety of other clinical applications have been found for alumina. These include knee prostheses, bone segment replacements, bone screws, middle ear bone substitutes, and corneal replacements. In dentistry, alumina has been used in various dental implants, including blade, screw, and post configurations. Alumina has also been used in jaw bone reconstruction. In addition, some dental implants have been fabricated from single-crystal Al_2O_3 (sapphire).

Table 1. Ceramic Biomaterials Categorized by Chemical Composition

Category	Example
Simple Oxides	Al_2O_3 and ZrO_2
Hydroxyapatite	HA Coatings on Hip Prostheses
Other Calcium Salts	Tricalcium Phosphate
Silicates	Bioglass
Glass-Ceramics	Apatite/Wollastonite (A/W)
Ceramic-Matrix Composites	A/W Glass-Ceramic with a Tetragonal Zirconia Dispersion

Because of its substantially higher fracture toughness, zirconia, ZrO_2, has become a popular alternative to alumina as a structural ceramic. [5] Zirconia has the largest value fracture toughness of any monolithic ceramic. Static and fatigue strengths for zirconia femoral heads have been found to exceed clinical requirements, and, more important, the decreased frictional torque leads to a reduced level of polyethylene debris production. [6] Wear performance is superior even to alumina which is superior to that of metal alloys. In general, the low wear rate of both alumina and zirconia in comparison to metal alloy heads produces negligibly small amounts of metal ion release. [7] Zirconia heads, because of their low modulus and high strength, can be manufactured in a greater range of sizes and neck lengths. On the other hand, stress raisers (high points on the conical stem) must be avoided to prevent fracture of the ceramic head at extremely low loads.

As we have seen, alumina had become the bioceramic of choice by the end of the 1970's based on its combination of biocompatibility and strength. The subsequent development of relatively high-fracture toughness zirconia ceramics has expanded the potential applications of these "simple" oxides. The most significant area of growth for bioceramics, however, involves the more complex material, hydroxyapatite.

It is perhaps ironic that such an obvious candidate as *hydroxyapatite* was not widely used as a biomaterial for many years. Hydroxyapatite, $Ca_{10}(PO_4)_6(OH)_2$, is the primary mineral content of bone representing 43 % by weight. It has the physiochemical advantages of stability, inertness, and biocompatibility. Its relatively low strength and toughness, however, produced little interest among researchers searching for bulk structural materials. The recent widespread and successful application of hydroxyapatite has largely been in a thin, surface-reactive coating applied to a variety of prosthetic implants, primarily for total hip replacement. [8] These coatings have been plasma-sprayed on both Co-Cr and Ti-6Al-4V alloys, with optimal performance coming from thicknesses on the order of 25-30 micrometers. Interfacial strengths between the implant and bone are as much as 5 to 7 times as great as with the uncoated specimens, corresponding to the mineralization of bone directly onto the hydroxyapatite surface with no signs of intermediate, fibrous tissue layers. The substantial success of this coating system has led to its widespread use in total hip replacement prostheses. One can expect this application to be expanded with increasing clinical experience and follow-up research and development activity.

Another successful application of a hydroxyapatite-containing biomaterial involves a range of commercial products for the purpose of repairing large bone defects, defined as centimeter-scale gaps in the skeletal system. Historically, such defects have been repaired by harvesting bone from another part of the body (autogenous bone grafting or *"autografts"*) or using cadaver bone (*"allografts"*). The harvest of an autogenous bone graft carries significant morbidity and cost. [9] Allografts have problems with immunologic reaction and the risk of acquiring diseases transmissible by tissues and fluids. These limitations have created substantial interest in the development of materials as bone graft substitutes or extenders. A pioneering example of hydroxyapatite in conjunction with tricalcium phosphate for the repair of large bone defects was the use of biphasic granules of these ceramics in a matrix of collagen, as described by McIntyre, et al. [10] An alternative material is the so-called "coralline hydroxyapatite," manufactured from coral by a thermochemical process which converts the calcium carbonate manufactured by the marine organism to a calcium phosphate (hydroxyapatite). An attractive feature of using the coral-route is that it has a "natural" open porous structure (greater than 100 μm) which is ideal for accommodating bone ingrowth.

A recent, comprehensive study sponsored by the Department of Veterans Affairs provided a comparison of a variety of commercially available ceramics as bone graft substitutes. [11] Cancellous (or spongy) bone harvested from the patient's iliac crest was taken as a standard for comparison with candidate synthetic materials. This study compared three commercially available granular ceramic materials consisting of hydroxyapatite and *other calcium salts*: a coralline-based hydroxyapatite (Interpore International, Irvine, CA), a β-tricalcium phosphate (DePuy, Warsaw, IN), and the biphasic (HA/TCP) ceramic/collagen composite (Zimmer, Warsaw, IN). Evaluations were based on the performance of the ceramic in repairing a 25 mm defect in a canine radius bone, in comparison to the performance of a cancellous bone autograft in the opposite leg of the same dog. All three ceramic samples were tested with and without the addition of bone marrow. The study found that HA and TCP alone are unsuitable substitutes. The addition of bone marrow made these ceramics comparable in performance to the cancellous bone graft after six months of implantation, apparently due to cells provided by the bone marrow imparting the stimulation of bone growth at the ceramic surface within the first month of implantation. The ceramic/collagen composite demonstrated competitive performance with or without bone marrow addition, although the performance may be maximized by additional marrow. It appears that collagen may facilitate bone formation by serving as a requisite to forming endochondral bone (i.e., the "long bones" of the skeleton). The authors of the study, however, preferred TCP with bone marrow to the ceramic/collagen material because HA, unlike TCP, is not readily resorbed by the body and is more opaque radiographically than TCP making radiographic evaluation of the degree of healing and bone formation difficult. For this reason, the authors suggest that an "ideal" graft material might be TCP plus collagen and bone marrow.

While *silicates* are a dominant part of the traditional ceramics and glass industries, [5] they have played a less significant role as bioceramics. The specialized requirements of biomedical applications have proved more important than the economics of widely available raw materials. Biomedical applications for crystalline silicates have been especially limited. For noncrystalline silicate glass, however, biomedical applications have been more significant because of the development of *Bioglass*, the classic "surface reactive" material as introduced in the opening, historical chapter and discussed thoroughly in the chapters by Hench and Kokubo. [12, 13] We have seen in numerous examples from both Hench and Kokubo that Bioglass is a "bioactive" material, able to bond to bone and, with specialized compositions, even to soft tissues.

An important variety of crystalline ceramics are the *"glass-ceramics,"* first produced like ordinary glassware and then transformed into crystalline ceramics by heat treatment. The benefit of the initial, glass stage is the ability to form the product into a complex shape economically and precisely. The benefit of the subsequent crystallization step is a final, fine-grained microstructure with little or no residual porosity, providing good resistance to mechanical shock due to the elimination of stress-concentrating pores. The crystallization process is not always 100% complete, and the residual glass phase effectively fills the grain boundary volume, helping to create the pore-free structure. Conventional glass ceramics are based on compositional systems such as $Li_2O-Al_2O_3-SiO_2$, which produce crystalline phases with exceptionally low thermal expansion coefficients and subsequent resistance to thermal shock. Glass-ceramics for biomedical applications are more often based on compositions similar to Bioglass. Conveniently, P_2O_5 serves as a nucleating agent, eliminating the need for TiO_2 additions used in conventional glass-ceramics. Low-alkali (0 to 5 wt%) silica glass-ceramics, known as *Ceravital*, have been successfully used for more than a decade as implants in middle-ear surgery to replace bone damaged by chronic infection. [4] In Japan, a two-phase silica-phosphate material known as A/W glass-ceramic, has been produced consisting of an apatite phase, $Ca_{10}(PO_4)_6(OH_1F_2)$, a wollastonite phase, $CaO \bullet SiO_2$, and a residual glassy matrix. It has been used successfully in hundreds of patients for replacing part of the pelvic bone and in vertebral surgery. In Germany, an easy-to-machine silica-phosphate glass-ceramic has been developed containing phlogopite (a type of mica) and apatite crystals. The compositional ranges in which bioglasses and bioglass-ceramics bond effectively with bone and other tissues is generally limited. Care is required. For example, small amounts of Al_2O_3 and TiO_2 can inhibit bone bonding.

As noted above, zirconia ceramics are popular alternatives to alumina because of relatively high fracture toughness values. *Ceramic-matrix composites* (CMC's) can have even higher values, comparable to some common structural metal alloys. (See Table 2.) In CMC's, mechanisms such as fiber pull-out cause crack growth to be retarded giving the higher fracture toughness values. As with silicate ceramics and glasses, the more common CMC's used in industry are not necessarily appropriate for biomedical applications. In addition to the advantage of improved fracture toughness, design goals in developing CMC's for biomedicine have focused on increasing flexural strength and strain to failure, while decreasing elastic modulus. [4] A good example is an A/W glass-ceramic containing a dispersion of tetragonal zirconia which has a bend strength of 703 MPa and a fracture toughness of 4 $MPa \bullet m^{1/2}$.

Table 2. Comparison of the Fracture Toughness of Ceramic Matrix Composites (CMC's) with Other Structural Materials

Material	K_{IC} $(MPa \bullet m^{1/2})$
CMC's	
SiC fibers in SiC	25
SiC whiskers in Al_2O_3	8.7
Other materials	
Pressure vessel steels	170
Aluminum alloys (high to low strength)	23-45
Partially stabilized zirconia	9
Sintered alumina	3-5
High-density polyethylene	2
Silicate glass	<1

FUTURE TRENDS IN APPLICATIONS OF BIOCERAMICS

In the previous section, we outlined the broad menu of ceramics that are currently finding applications in biology and medicine. Now, we can survey future trends in applications for these *bioceramics*. Our focus shall be on three areas, viz. orthopaedics, dentistry, and cancer treatment.

Within *orthopaedics*, our initial focus will be total hip replacement (THR) surgery, a highly successful and widely used procedure, with more than 200,000 THR surgeries performed in the

United States each year and a similar number in Europe. [14]. The THR was developed in England by Sir John Charnley, who was knighted for his achievement. The essence of his invention was to provide adequate fixation for the prosthesis used to replace the natural ball-and-socket of the hip joint (often defective due to degenerative arthritis). A metallic femoral stem is placed in an opening drilled into the medullary canal of the femur. Stainless steel was the original alloy used for both the stem and the attached ball. The cup on the acetabular side of the joint (in the hipbone) was fabricated of ultrahigh molecular weight polyethylene (UHMWPE). Charnley adapted, from dentistry, the use of *polymethylmethacrylate (PMMA)* cement to fix the femoral stem and acetabular cup. In the 1970's, the use of a porous surface on the metallic stem allowed fixation by bone ingrowth, a cementless alternative to PMMA. As with the development of bioceramics [1], materials selection for the THR was reasonably well established by the mid-1970s. [15] Stainless steel has largely been replaced by cobalt-chrome alloys (for cemented implants) and Ti-6Al-4V alloy (for cementless implants). The titanium alloy is undesirable for the cemented design due to its lower elastic modulus which leads to an excessive load on the interfacial cement. Conversely, the lower modulus of the titanium alloy makes it preferable for the cementless design, as less modulus mismatch with bone creates less stress shielding of the bone. Table 3 summarizes the common materials-of-choice for contemporary THR surgeries and shows potential ceramic alternatives to metals and polymers. Only in the past decade has a ceramic, *hydroxyapatite* for the cementless design, appeared among the contemporary choices. Its substantial success makes it an increasingly popular alternative in cementless designs. It should be noted that hydroxyapatite coatings do not have to be porous in order to provide strong bonding to bone. These plasma-sprayed coatings exhibit enhanced interfacial strength due to mineralization of bone directly onto the surface.

Table 3. Current Engineered Materials for the Total Hip Replacement (THR) and Potential Engineered Ceramic Alternatives

Component	Current Materials	Ceramic Alternatives
Femoral Stem	Co-Cr alloys or Ti-6Al-4V	Partially-stabilized zirconia (PSZ) or Ceramic-matrix-composites (CMC)
Ball	Co-Cr alloys	Al_2O_3 or PSZ
Acetabular Cup	Ultrahigh molecular weight polyethylene (UHMWPE)	Al_2O_3 or PSZ
Cement	Polymethylmethacrylate (PMMA)	A/W Glass-ceramic
Cementless	Porous surface coating or hydroxyapatite coating	Hydroxyapatite coating

The most challenging substitution would be a replacement for the metallic femoral stem. Polymer matrix composites are more probable competitors. [16] The compressive load on the femoral ball makes alumina and zirconia ceramics good candidates for that application, and we saw, in the previous section, that, at least in Europe, they have been used extensively for this purpose. An attractive feature of ceramics for femoral heads and acetabular cups is the typically low surface wear of structural ceramics. The low toughness of these materials, however, has contributed to a continued dominance by polyethylene. Recently, Yamamuro, Kokubo, and co-workers have developed an attractive ceramic alternative to PMMA cement. [17] Based on the A/W glass-ceramic mentioned earlier, their cement has similar behavior to PMMA, with the advantages of an adherent bond with bone and an absence of a large exothermic reaction (which prevents the building of bone cells at the cement interface).

In general, joint replacement surgeries are considered increasingly difficult as one moves radially away from the hip joint. Closest to the hip, the knee has been the joint which has benefited most from the technology of the THR surgery, and knee replacement surgery is now widely practiced in the United States. Most prostheses are similar to THR's in materials selection, with a femoral component made of either cobalt-chrome or titanium alloys and a high-density polyethylene (HDPE) for the wear surface connected to the tibia bone, often with a metal backing. The knee prosthesis also includes a patella (knee cap), also made of HDPE but without a metal backing. Because of a geometry very different than the THR, there is no particular advantage of cobalt-chrome versus titanium alloy for the femoral component of the knee prosthesis. The trade-off between the use of PMMA cement versus cementless fixation is similar to that for the THR. Once the art of prosthesis alignment was mastered, cement fixation proved to be as successful as cementless. Hydroxyapatite coatings for prosthesis fixation should have substantial promise in knee replacement surgery, although HA applications in hip surgery are more widely used at this time. The replacement of other joints such as the shoulder, elbow, and especially the ankle are substantially greater biomechanical challenges. In the more limited and experimental surgeries for the replacement of these joints, ceramic applications should not be expected to provide a serious challenge in the near future to the common use of metals and polymers.

We saw in the previous section that ceramics are widely used for bone defect and fracture repair. Novel new bioceramics are being developed for this area of application. One example is an *in situ* ceramic processing technique that produces a more natural hydroxyapatite (HA) by the surgical implantation of a paste that hardens in minutes under physiological conditions. [18] (The HA formed by conventional ceramic processing techniques tends to be more dense, coarse-grained, and less fatigue-resistant than HA formed *in vivo*.) The novel paste is produced by adding a sodium phosphate solution to a mixture of monocalcium phosphate monohydrate [MCPM, $Ca(H_2PO_4)_2 \cdot H_2O$], α-tricalcium phosphate [TCP, $Ca_3(PO_4)_2$] and calcium carbonate (CC, $CaCO_3$). Once set, the material's tensile strength is $\approx$2.1 MPa, about the same as that for cancellous bone. The average grain size for the crystallized paste is $\approx$20-50 nm, again comparable to that in natural bone. Using this paste, fractured bones can be held in place while natural bone remodeling occurs, replacing the implant with an implant-bone composite. This bioceramic is marketed under the name *Norian SRS* (for "skeletal repair system"), and mimics the mineralization process of coral, which involves physiochemically controlled reactions. In general, the mineralization of bone involves protein-directed reactions, a process that is only recently being controlled clinically. The utility of bone marrow, as noted earlier, for stimulating bone growth appears to be functioning in the same way. A more sophisticated approach is the use of bone morphogenetic proteins (BMP), whose therapeutic effectiveness for bone regeneration is now well established. [19] The clinical effectiveness of BMP can be enhanced by a delivery system, and a leading candidate in this regard is porous hydroxyapatite. [20] This use of HA ceramics as delivery systems for bone inductive proteins may represent the cutting edge of bioceramic applications.

Given the relatively small scale and primarily compressive loads involved, a wide variety of ceramics, glasses, and glass-ceramics have been applied to *dentistry*. [4, 21, 22] Examples include dental implants, cementation agents, and restorative materials. *Dental porcelains* have been used for nearly two centuries to repair diseased and decayed teeth, [22] with approximately 80% of all fixed prostheses placed in the United States being porcelain-fused-to-metal. Porcelains are also used for the cosmetic treatment of broken and discolored teeth. These porcelains are generally made by blending two components, such as partially fused potassium aluminosilicate feldspar (leucite) and fully fused alkali - alkaline-earth - aluminosilicate. Additives provide the desired optical properties, including color and fluorescence. A typical metal substrate is an 80Ni-20Cr alloy. Good adhesion of the porcelain depends on good thermal expansion matching as well as a preoxidation step to produce a monolayer of Cr_2O_3. As pointed out in the previous section, relatively pure alumina, including the single crystal sapphire form, has been used as dental implants. Alumina has also been used in jaw bone reconstruction.

Earlier, we saw that some of the primary applications of *Bioglass* implants are in the dental field, [22] including implantation in the jaw bone to fill the cavity caused by removal of teeth due to disease or injury. Well over 90% of such implants are successful five years beyond implantation, compared with losses greater than 50% for implants made from other materials. Thousands of patients with periodontal disease are treated with a mixture of crushed Bioglass particles (90 to 700 nm) and saline placed around the tooth to stimulate bone growth. One should also note the article by Hench in this volume. [12]

Dental cements are composed of a variety of complex chemical systems, [21] many involving ceramics. Often the cement functions by a mechanism of mechanical locking, rather than true chemical adhesion. Resin-composites have been increasingly used to repair and rebuild teeth since being introduced into dentistry in the early 1960's. [21] These systems consist of an organic polymer (resin) matrix, a ceramic filler (typically quartz, colloidal silica, or silicate glasses containing strontium or barium), and a coupling agent (binder) between the two components. The popularity of the resin-composites is based on their more desirable aesthetics compared to metal amalgams, as well as a growing concern about the potential danger of mercury in traditional amalgams. An especially interesting comparison of bioceramics with organic and metallic dental materials is given in this volume in the paper by Tweden, et al. [23]

An exciting new area for the application of bioceramics is *cancer treatment*. An emerging technique for *in situ* irradiation involves the dissolution of a β-emitting radioisotope in chemically insoluble, biocompatible, and nontoxic glass microspheres. [22,24] They must be insoluble to ensure that the radioactive material is not released into the body. The microspheres must be sized to lodge in the capillary bed of the organ to be treated. Yttria aluminosilicate (YAS) glasses meet these design criteria. Preliminary clinical studies on liver-cancer patients have been highly promising, with significantly longer survival times than with conventional treatment. Doses as high as 15,000 rads can be safely delivered by this technique, with minimum side effects providing improved quality of life for the patients. An alternate approach to cancer treatment using bioceramics is the incorporation of a ferromagnetic phase which allows the killing of cancer cells by local heating. Kokubo [25] has described the development of a glass-ceramic which is both bioactive and ferromagnetic. Kokubo discusses these various approaches to cancer therapy in his article in this volume. [13]

Finally, we should note that the previous chapter from Professor Sarikaya focuses on ways in which studies of biomaterials have led to new concepts for producing engineered materials. [26] This fertile area of research and development is highly synergistic, as new materials produced in these novel ways which imitate natural, biological processes may prove to be superior candidates for biomedical applications. *Biomimetic processing* is an outgrowth of the concentrated effort about two decades ago in the fabrication of ceramics and glasses by *sol-gel processing*, in which this "chemical route" to manufacturing could be done, for a given material, at a substantially lower temperature than by conventional processing. The efforts to produce biomimetic materials, by the merging of materials science with the biological sciences, can lead to processing techniques near ambient temperature. These imitate natural processes of low-temperature aqueous syntheses of oxides, sulfides, and other ceramics. [27] A good example of biomimetic processing is a novel method for producing ceramic thin films by precipitation from aqueous solution, allowing ceramic films to be applied on surfaces which are not amenable to conventional ceramic coating techniques. [28]

References

1) Shackelford, J.F.: *this volume*, 1.
2) Shackelford, J.F.: *Bioceramics*, 1998, Gordon and Breach Science Publishers, Amsterdam.
3) Hulbert, S.F., Hench, L.L., Forbers, D., and Bowman, L.S.: *Ceramurgia Intl.*, 1982-83, **8-9**, 131.
4) Hench, L.L.: *Bull.Amer.Ceram.Soc.*, 1993, **72**, 93.
5) Shackelford, J.F.: *Introduction to Materials Science for Engineers*, 4th Edition, 1996, Prentice-Hall, Upper Saddle River, NJ.
6) Kumar, P., Masanori, O., Ikeuchi, K., Shimizu, K., Yamamuro, T., Okumura, H., and Kotoura, Y.: *J.Biomed.Mater.Res.*, 1991, **25**, 813.
7) Davidson, J. and Kovacs, P.: *Smith & Nephew Richards Research Report, OR-90-6*.
8) de Lange, G.L. and Donath, K.: *Biomaterials*, 1989, **10**, 121.
9) Younger, E.M. and Chapman, M.W.: *J. Orthopaedic Trauma*, 1989, **3**, 192.

10) McIntyre, J.P., Shackelford, J.F., Chapman, M.W., and Pool, R.R.: *Am.Ceram.Soc.Bull.*, 1991, **70,** 1499.

11) Johnson, K.D., et al., *J. Orthopaedic Research*, 1996, **14**, 351.

12) Hench, L.L.: *this volume*, 37.

13) Kokubo, T.: *this volume*, 65.

14) *Operative Orthopaedics*, Second Edition, Vols. 1-4, ed. Chapman, M.W.: 1993, J.P. Lippincott, Philadelphia.

15) Fraker, A.C. and Ruff, A.W.: *J. Metals*, 1977, **29**, 22.

16) Hunt, M.S.: *Materials and Design,* 1987, **8,** 113.

17) Kawanabe, K., Yamamuro, T., Nakamura, T., Kokubo, T., Yoshihara, S., and Shibuya, T.: p. 233 in *Bioceramics 5*, eds. Yamamuro, T., Kokubo, T., and Nakamura, T.: 1992, Kobonshi Kankokai, Kyoto, Japan.

18) Constantz, B.R., et al.: *Science*, 1995, **267**, 1796.

19) Reddi, A.H.: *Cytokine & Growth Factor Reviews*, 1997, **8**, 11.

20) Ohgushi, H., Okumura, M., Yoshiko, D., Shiro, T., and Tamai, S.: to be published in *Proc. Second Intl. Conf. on Bone Morphogenetic Proteins*, ed. Reddi, A.H.: Sacramento, CA, June 4-8, 1997.

21) Clark, A.E. and Anusavice, K.J.: p. 1091 in *Engineered Materials Handbook*, Vol. 4, 1991, ASM International, Materials Park, Ohio.

22) Day, D.: *Am.Ceram.Soc.Bull.*, 1995, **74,** 64.

23) Tweden, K.S., Maze, G.I., McGee, T.D., Runyon, C.L., and Niyo, Y.: *this volume*, 17.

24) White, J.E. and Day, D.E.: p. 181 in *Key Engineering Materials*, Vol. 94-95, 1994, TransTech, Cleveland, Ohio.

25) Ikenaga, M., et al.: p. 255 in *Bioceramics*, Vol. 4, 1991, Butterworth-Heinemann, Guilford.

26) Sarikaya, M.: *this volume*, 83.

27) Heuer, A.H., et al., *Science*, 1992, **255,** 1098.

28) Bunker, B.C., et al.: *Science*, 1994, **264,** 48.

AUTHOR INDEX

KEYWORD INDEX

Porous Ceramic Materials
Fabrication, Characterization, Applications

Ed. Dean-Mo Liu

Key Engineering Materials Vol. 115

ISBN 0-87849-706-4
1996, 248 pp, SFr 120.00 (ca. US$ 92.00)

The development of porous ceramic materials has brought a new challenge to a variety of industries because porous ceramics are more durable in severe environments and their surface characteristics permit them to satisfy specific functional purposes. With the growing demands of porous ceramics for industrial applications, a number of technologies have been developed to fabricate these materials with an attempt to control their pore characters as well as to realize the pore-related properties in order to gain a deeper understanding of the relation between the various pore-related properties for optimization purposes. To date, porous ceramic materials with more delicate and uniform pore structures and pore sizes ranging from a few hundred micrometers to a few nanometers can be achieved for diverse purposes by either physical or chemical processing.

It is one of the purposes of this special volume to bring readers some or better understanding of the processing and properties of the porous ceramic materials. It presents a collection of papers covering the fabrication, property evaluation, characterization and applications of the porous ceramic materials developed to date. Included are fundamental theories, novel fabrication techniques, and special classes of ceramic materials involved in sensing and biomedical applications.

List of Contributors:

H. Abe, R.B. Bagwell, C. Cantalini, W.J. Chao, H.I. Chen, W.Q. Chen, A.S.T. Chiang, K.S. Chou, M. Egashira, A Fukunaga, D.A. Hirschfeld, T.C. Huang, L.C. Klein, C.K. Lee, T.K. Li, F.H. Lin, C.C. Lin, D.M. Liu, H.C. Liu, G.L. Messing, S. Morimoto, K.S. Patel, M. Pelino, S.X. Qu, R.W. Rice, H.T. Sun, H. Tateyama, C.S. Tsay, H. Tsuzuki, C.Y. Wang, R.H. Woodman, Z.J. Yang, T. Yazawa, M.Q. Yuan, M.Q. Zhang

Detailed information on this title – including the complete table of contents – is available on the internet at http://www.ttp.ch/titles/706.htm or through TTP's E-Mail Preview Service. For further information, please send an e-mail to 'preview@ttp.ch' with the word 'help' as the body of the message.

 **Trans Tech Publications Ltd**

Brandrain 6
CH-8707 Uetikon-Zuerich
Switzerland

Fax: +41 (1) 922 10 33
e-mail: ttp@ttp.ch
Web: http://www.ttp.ch

Your Direct Access

World Wide Web

Please visit us on the World Wide Web at

http://www.ttp.ch

where detailed information on all published titles is provided as well as:

- Online Shopping
- Site Search
- Full Tables of Contents
- New & Forthcoming Titles
- Periodicals & Book Series
- Preview Service
- Download Area

TTP Preview Service

Trans Tech Publications' preview service offers automatic delivery by e-mail of information on new books and periodical issues in *your area of interest*, including tables of contents - several weeks before the actual release of the respective publication.

Included are all titles published including the following periodicals and book series:

- Advanced Manufacturing Forum
- Advanced Materials Research
- Defect and Diffusion Forum
- Environmental Research Forum
- GeoResearch Forum
- Key Engineering Materials

- Materials Science Forum
- Materials Science Foundations
- Molten Salt Forum
- Production and Logistics Forum
- Solid State Phenomena

This service is free of charge. For details please send an e-mail message containing the line
help
to **preview@ttp.ch**
or register your e-mail address and areas of interest online at http://www.ttp.ch

For regular e-mail please use the address **ttp@ttp.ch**

ttp Trans Tech Publications Ltd

Brandrain 6 • CH-8707 Uetikon-Zuerich • Switzerland
Fax +41 (1) 922 10 33 • e-mail: ttp@ttp.ch
http://www.ttp.ch

Porous Materials for Tissue Engineering

Eds. Dean-Mo Liu and Vivek Dixit

Materials Science Forum Vol. 250

ISBN 0-87849-773-0
1997, 252 pp, CHF 125.00 / US$ 94.00

Tissue Engineering is an interdisciplinary field of study that involves the development of (bio)artificial organs and implants. The primary purpose of such prosthesis is repair, regeneration, and reconstruction of lost, damaged, or degenerative tissues/organs. This newly emerging scientific discipline is derived from the fusion of the principles of engineering and biology with the ultimate goal of creating fully functional biological substitutes. The unique scientific union has brought us to a new era of understanding in the field of biomedicine. It will enable us to restore lost function of human tissue and thus help to improve mankind's overall quality of life.

There is tremendous interest in this fascinating field and the present volume has been intended to provide further insight into the better understanding of the Tissue Engineering from a variety of scientific/engineering/biomedical viewpoints. The book presents a collection of articles concerning the development, biomaterial characterization, and evaluation of possible applications that are currently being investigated in tissue engineering.

SOME HIGHLIGHTS

Biomaterials in Different Forms for Tissue Engineering: An Overview
E. Pişkin

Design of Macroporous Biodegradable Polymer Scaffolds for Cell Transplantation
V. Maquet and R. Jerome

Synthetic Extracellular Matrices for Cell Transplantation
M.C. Peters and D.J. Mooney

Porous Hydrogels for Neural Tissue Engineering
S. Woerly

Substrates for Growth Cone Guidance in Brain: Guidance Cues for Neural Connections
K. Torimitsu

Detailed information on this title – including the complete table of contents – is available on the internet at http://www.ttp.ch/titles/773.htm or via TTP's E-Mail Preview Service. For further information, please send an e-mail to 'preview@ttp.ch' with the word 'help' as the body of the message.

 **Trans Tech Publications Ltd**

Brandrain 6
CH-8707 Uetikon-Zuerich
Switzerland

Fax: +41 (1) 922 10 33
e-mail: ttp@ttp.ch
Web: http://www.ttp.ch